Felix Bloch

RICE UNIVERSITY STUDIES

FELIX BLOCH AND TWENTIETH-CENTURY PHYSICS

Dedicated to Felix Bloch
on the Occasion of his Seventy-Fifth Birthday

M. CHODOROW, R. HOFSTADTER, H. E. RORSCHACH, and A. L. SCHAWLOW, Editors

H. Gutfreund
E. L. Hahn
Walter A. Harrison
Oleg Jardetzky
Carson D. Jeffries
Walter Kohn
L. Kowarski
Willis Lamb
W. A. Little
John D. Roberts
Michael Schick
Hans H. Staub
H. Suhl
Norma Wade-Jardetzky
J. D. Walecka
Chen Ning Yang
Avishay Yaniv

PUBLISHED BY
WILLIAM MARSH RICE UNIVERSITY
HOUSTON, TEXAS

Vol. 66, No. 3 Summer 1980

US-ISSN-0035-4996
US-ISBN-0-89263-246-1

This work relates to Department of the Navy Research Grant N00014-80-G-0063 issued by the Office of Naval Research.

FELIX BLOCH

This volume is dedicated to Felix Bloch on the occasion of his seventy-fifth birthday. The editors hope that these papers by former students and colleagues will entertain and enlighten and will convey a sense of the high respect and regard in which Felix is held by all who have been privileged to be his associates. We also hope that the broad range of these papers will be indicative of the impact of his work on twentieth-century physics.

One expects that book-editing will be a frustrating chore. This book was quite otherwise. The authors were eager to write, quick to submit their manuscripts, and anxious to be a part of this heartfelt expression of love and respect.

We editors want to express our own gratitude to Felix Bloch. As friends, associates, and colleagues for many years, we can presume to speak for all who came in contact with Felix. We have treasured his friendship, been inspired by his rigorous insights, and been brought to a greater love for and devotion to the beauty of physics by his example. It is with profound gratitude and love that we present this volume to him.

"I always felt Bohr was somewhat disappointed that I was not more interested in epistemology, but being of a pragmatic nature, I always wanted to deal with concrete problems." Felix Bloch spent six months with Niels Bohr in the winter of 1931-1932. Even then, at the age of 26, he was a well-respected theorist and had come into contact with many of the prominent European physicists who were associated with the development of quantum mechanics. Bloch had first met Bohr in 1929, when he had gone with H. A. Kramers to one of Bohr's meetings held annually in Copenhagen. "I learned 'qualitative' physics from him," recalls Bloch. Those who have heard Bohr speak will know that he was not easy to understand. But a part of his inspiration was due to this very characteristic. Bloch found conversations with Bohr to be often "confusing," but they left him with an urge to clarify the physics by a detailed calculation.

Many of the European physicists of that period owe their initiation in modern physics to Arnold Sommerfeld, whose monumental work

Atombau und Spectralinien can be compared in its impact only with P. A. M. Dirac's "The Principles of Quantum Mechanics." Bloch was no exception. He learned modern physics from Sommerfeld's book, but that was not his only tie to Sommerfeld. He was the first student of Werner Heisenberg, who had in turn been Sommerfeld's most brilliant student. Bloch had begun his study of physics at the Federal Institute of Technology (ETH) in his home town of Zürich. He was strongly influenced by Peter Debye, having taken Debye's introductory physics course, in which he was greatly impressed by Debye's ability to catch the simple and essential point of a problem. When Debye left Zürich for Leipzig in 1927, he advised Bloch to come to Leipzig and to work there under Heisenberg. Bloch accepted this advice and after being introduced to Heisenberg told him about calculations on the radiation damping of wave packets that he had begun in Zürich. Heisenberg encouraged Bloch to complete this work, later published as his first paper.[1] He then suggested that the time was ripe for the fruitful application of quantum mechanics to the solid state and that the electric conduction of metals might be a good topic to consider. This work, which Bloch began immediately, was not only his thesis, but, when published in 1928,[2] established the theory of metals in its modern form and brought to Bloch a reputation as a brilliant and promising young theorist.

In 1928, Bloch returned to Zürich to be Wolfgang Pauli's assistant. Pauli proposed that he work on the theory of superconductivity, since in his opinion this was the only remaining problem of importance in solid-state physics. Bloch tried to explain persistent currents by the fact that they correspond to states at a relative minimum of the energy but under seemingly quite general conditions found that no such state carried a current. This result has often been referred to as "Bloch's Theorem" and was jocularly summarized by the statement that "any theory of superconductivity can be disproven." The statement is an oversimplification, but the result was an important contribution, indicating that the phenomenon of superconductivity was more profound than had been revealed by the existence of persistent currents. Bloch also published work[3] on the ferromagnetic behavior of free electrons, but it was not a very happy year for Felix, particularly since he had hurt a knee while skiing that winter.

The year that followed was a much happier one. In 1929 Bloch traveled to Holland as a Lorentz Fellow, where he spent six months with Kramers in Utrecht and a few months with A. D. Fokker in Haarlem. This was a fruitful time, during which he developed the T^5 law for the temperature dependence of the electrical resistance,[4] thus proving that the profound analysis of the electron-lattice interaction first introduced in his thesis was capable of explaining in detail the experimental results on the low-temperature resistance of metals. He also published his paper containing the introduction of spin waves,[5] which became the forerunner of all modern quasiparticle theories. (Bloch has often referred to this work as an "embellishment" on Heisenberg's theory of ferromagnetism!) Kramers was a delightful host, and he and Bloch became good friends. L. D. Landau, whom Bloch had met earlier at Copenhagen, also visited him in Utrecht, and their close friendship led Bloch to visit Landau later (in 1931) in Russia.

In 1930, Heisenberg asked Bloch to return to Leipzig as his assistant. It was at this time that Bloch's work on ferromagnetism was extended, culminating in his discovery of "Bloch Walls" which were to prove of great importance in understanding the magnetic properties of real materials, especially the phenomena of remanence and hysteresis. This work was included in his "Habilitationsschrift" on the exchange problem in magnetism.[6] (This is a kind of second thesis, required of those who wish to pursue an academic career at a German university.)

After one year with Heisenberg, Bloch was invited to Copenhagen for six months as an Oersted fellow in Bohr's institute. He established a close friendship with Bohr during this period—a friendship that was to last through the trying years of the war and for the rest of Bohr's life, and one that would be renewed by frequent visits. (Bloch has remarked that he was in almost continual contact with Bohr through the years, and that the same deep problems never ceased to occupy Bohr's mind.) As a result of conversations with him, Bloch became interested in the stopping power of matter for charged particles,[7] a subject in which Bohr had had a long interest.

In 1932, Bloch returned to Leipzig, where he had become a lecturer and continued as Heisenberg's asistant. In January 1933, Hitler was appointed Chancellor of Germany. This event was to have the same profound effect on Bloch's career as it did on such other great physicists as Albert Einstein, James Franck, and Hans Bethe. Bloch realized that he would have to leave Germany, and he spent the spring and summer of 1933 in Switzerland and in Paris (where he stayed at the home of Paul Langevin). He had applied for and was awarded a Rockefeller Fellowship for a year, starting in the fall of 1933, which he intended to use partly in Rome with Enrico Fermi and partly at the Cavendish Laboratory in Cambridge. On his way to Rome, he visited Bohr at Copenhagen, where he received a letter from the Chairman of the Stanford Physics Department, David L. Webster, offering him a position at Stanford University. He knew nothing of Stanford, and he turned for advice to Heisenberg, who was also visiting Bohr. Heisenberg remarked, "O yes, those are the people who steal one another's axes" (a reference to the Stanford-Berkeley football rivalry). Bohr was more helpful, since he had met Webster and some of the other faculty at Stanford and knew that it was a fine university. Bloch continued on to Rome, but Bohr's advice persuaded him to accept Webster's offer and to abandon his planned visit to Cambridge.

Bloch's time in Rome was short, but he learned something of Fermi's style. "Fermi was not stimulating in the sense of suggesting to you a new problem," Bloch recalls. "But he would always insist on absolute clarity in any discussion. He also said to me as I was preparing to leave Rome: 'You really should start to do some experiments. They are fun!' "

This proved to be good advice. With Fermi as an outstanding exception, most European physicists of those days were sharply divided into theorists and experimentalists. The experimentalists often did not know much theory and experimentation was regarded by many theorists to be below their dignity. When Bloch arrived at Stanford in 1934, he began a close collaboration with the experimental work of N. E. Bradbury on gas discharges and with that of P. A. Ross on x rays.[8] Robert Oppenheimer was at Berkeley, and Bloch had already met him in Zürich while serving as Pauli's assistant. Oppenheimer organized a joint Stanford-Berkeley weekly seminar, which Bloch regularly attended. Quantum electrodynamics was the frontier of theoretical physics then, and Bloch wrote (with A. Nordsieck) a paper on the infrared catastrophe,[9] which Heisenberg always regarded to be of particular importance.

While Bloch was still in Leipzig, Heisenberg once told him his ideas about nuclei while they were walking together. The neutron had been discovered in 1932, and Heisenberg and E. Majorana had proposed that the elementary constituents of the nucleus were protons and neutrons. Heisenberg's ideas had stimulated Bloch's interest in neutrons, and he returned from a visit to Europe in 1935 with the resolve that Stanford should begin experimental work on neutron physics. Bloch had found that the experimental physicists at Stanford knew much more theory than their European counterparts, and that it would be easy and exciting to follow Fermi's advice. This turned out to be a crucial point in Bloch's career: The neutron work combined with his earlier work on magnetism set him on the path leading to his discoveries in the field of nuclear magnetism, for which he and E. M. Purcell were awarded the Nobel Prize in physics in 1952.

The work in neutron physics proved to be extremely fruitful, and this period is described in some detail in an article in this volume by H. Staub. The most important papers of this period were those on the magnetic scattering of neutrons.[10] Of all his papers, Bloch is proudest of these, since they bring together two widely diverse phenomena: the magnetic properties of materials and those of the neutron. To Bloch one of the important features of the creative enterprise is this kind of unification (to reveal the common features in two apparently quite different fields of physics), and he ascribes these important discoveries to his familiarity with both magnetism and neutron physics. The impact of this work was great. It demonstrated the principle of a method used for the production of polarized neutrons and stimulated a new line of research, which has led to much of our modern understanding of the structure and dynamics of magnetic materials through the powerful technique of neutron scattering.

At an April meeting of the American Physical Society in 1939, Bloch met Lore Misch at the New York home of a friend to whom he had been introduced by F. Rasetti. Lore had been born in Berlin and was a physicist trained in x ray crystallography. She had left Europe to escape Hitler in 1938 and was working in George Harrison's spectroscopy laboratory at M.I.T. In September 1939, Bloch spent five weeks in the East and at that time he and Lore decided to get married. The marriage took place in Las Vegas the next spring. They had four children, and the Blochs have been a happy and stable family.

The war years brought Manhattan Project work to Stanford, where Bloch remained doing research on uranium in collaboration with Staub and in close contact with Oppenheimer until 1943. In the summer of 1943, Bloch went to Los Alamos, but, at the request of J. H. Van Vleck, he departed after a few months for the Radio Research Laboratory in Cambridge, where he was involved in antiradar work[11] until the end of the war. He returned to Stanford in summer 1945, still unaware of the progress on the development of an atom bomb and the decision to use the bomb in Japan.

Bloch's knowledge of magnetism and his work before the war on the magnetic properties of neutrons were now brought together in one of his most important discoveries—nuclear induction. Not only did he lay the theoretical groundwork[12] for the understanding of nuclear induction and its extension into many other fields, but also, with W. W. Hansen and M. Packard,[13] he made the first observation of proton resonance in water.

Bloch's work has had a great impact on further developments. His thesis work on electrical conductivity established the framework not only for understanding the conduction of electricity in metals but for all phenomena that involve the interaction of electrons with the crystal lattice. His discovery of the magnetic scattering of neutrons proved to be of great practical value for the study of the properties of solids by neutron scattering techniques. The fundamental importance of his work on nuclear magnetism was recognized when he shared the Nobel Prize for physics in 1952. The Nobel Committee could not have foreseen the tremendous influence that the idea of nuclear induction would have in physics, chemistry, biology, and even optics. Some of this influence is reflected in papers in this volume. The principles of nuclear induction have been enormously fruitful in their application to research on the structure and dynamics of chemical systems, and a whole industry has been built up around them.

Bloch considers himself to be an "applied" physicist. That is, he applies quantum mechanics to investigations in physics. (This is in contrast, he remarks, to Heisenberg, who "invented quantum mechanics.") Following his discovery of nuclear induction in 1946, he embarked on a program to "apply" nuclear induction, and he and his students began a series of experiments on the precise measurement of nuclear magnetic moments. Earlier (in 1940 with L. W. Alvarez), Bloch had made an important measurement of the moment of the neutron,[14] and he now measured the moments of the proton, deuteron, and triton.[15]

During the next decade, he and his students made many important theoretical and experimental contributions to the application of this powerful technique to other problems in physics. Bloch's profound insight and inspiring example during this period are described in articles in this volume by two of his collaborators, C. D. Jeffries and E. L. Hahn.

Bloch's importance to the physics of the twentieth century extends far beyond his published work. His deep understanding of physics and his firm but kind insistence upon rigor made a great and lasting impression on his students and colleagues. His influence extended to other fields beyond his own interests, and the article in this volume by W. Lamb illustrates this with an account of Bloch's influence on Lamb's own research. Bloch's stature in his field and the enormous respect in which he is held in the scientific community have led to many honors in addition to the Nobel Prize. In 1954-55 he served as the first Director-General of CERN; this period is reviewed in this volume in an article by L. Kowarski. He was elected to the National Academy of Sciences in 1948, and served as President of the American Physical Society in 1965. He received honorary degrees from the University of Grenoble, Oxford University, the Hebrew University of Jerusalem, the University of Zürich, Gustavus Adolphus College, Brandeis University, and the University of Pavia, and was awarded the German Order "Pour le Mérite" for Arts and Sciences. He was elected an honorary fellow of the Weizmann Institute of Science in 1958, an honorary member of the Royal Dutch Academy of Sciences in 1965, a fellow of the Royal Society of Edinburgh in 1966, and an honorary member of the French Physical Society in 1970.

Bloch's life in this country was centered at Stanford University, where, with Leonard Schiff, he built the physics department into one of the most prestigious in the world. He retired from his position as Max H. Stein Professor of Physics in 1971, but he has remained vigorous and active and continues as an inspiring colleague in the community of physicists.

THE EDITORS
M. CHODOROW
R. HOFSTADTER
H. RORSCHACH
A. SCHAWLOW

ACKNOWLEDGMENTS

The editors have received valuable editorial assistance from Professors C. M. Class and I. M. Duck of Rice University, from Professor M. R. Willcott of the University of Houston, from Kathleen Much, associate editor of RICE UNIVERSITY STUDIES, and from M. M. Comerford, G. D. Hale, and E. B. Truan of the Physics Department at Rice University. We acknowledge with gratitude the financial assistance of the Office of Naval Research.

REFERENCES

1. "Radiation Damping in Quantum Mechanics," Phys. Zeits. **29**, 58 (1928).
2. "Quantum Mechanics of Electrons in Crystal Lattices," Zeits. f. Physik **52.7-8**, 555 (1928).
3. "Electronic Theory of Ferromagnetism and Electric Conductivity," Zeits. f. Physik **57.7-8**, 545 (1929).
4. "Temperature Variation of Electrical Resistance at Low Temperatures," Zeits. f. Physik **59.3-4**, 208 (1930).
5. "Theory of Ferromagnetism," Zeits. f. Physik **61.3-4**, 206 (1930).
6. "Theory of the Exchange Problem and of Residual Ferromagnetism," Zeits. f. Physik **74.5-6**, 295 (1932).
7. "Stopping Power of Matter for Swiftly Moving Charged Particles," Ann. d. Physik **16.3**, 285 (1933); "Stopping Power of Atoms with Several Electrons," Zeits. f. Physik **81.5-6**, 363 (1933).
8. "Radiative Auger Effect" (with P. A. Ross), Phys. Rev. **47**, 187 (1935); "Mechanism of Unimolecular Electron Capture" (with N. E. Bradbury), Phys. Rev. **48**, 689 (1935).
9. "Radiation Field of the Electron" (with A. Nordsieck), Phys. Rev. **52**, 54 (1937).
10. "On the Scattering of Neutrons," Phys. Rev. **50**, 259 (L) (1936); "On the Scattering of Neutrons. II," Phys. Rev. **51**, 994 (L) (1937).
11. "Radar Reflections from Long Conductors," J. Appl. Phys. **17**, 1015 (1946).
12. "Nuclear Induction," Phys. Rev. **70**, 460 (1946).
13. "The Nuclear Induction Experiment" (with W. W. Hansen and M. Packard), Phys. Rev. **70**, 474 (1946).
14. "Quantitative Determination of the Neutron Moment in Absolute Nuclear Magnetons" (with L. W. Alvarez), Phys. Rev. **57**, 111; **57**, 352 (a) (1940).
15. "Relative Moments of H_1 and H_3" (with Graves, Packard, and Spence), Phys. Rev. **71**, 551 (L) (1947); "Relative Nuclear Moments of H^1 and H^2" (with E. C. Levinthal and M. E. Packard), Phys. Rev. **72**, 1125 (L) (1947).

RICE UNIVERSITY STUDIES

Vol. 66, No. 3 Summer 1980

FELIX BLOCH AND TWENTIETH-CENTURY PHYSICS

Dedicated to Felix Bloch
on the Occasion of his Seventy-Fifth Birthday

ORDERING AND IMPURITIES IN ONE-DIMENSIONAL METALS

by H. Gutfreund and W. A. Little

ABSTRACT

Using an extension of Nambu's formalism to handle singlet superconductivity, triplet superconductivity, charge density waves, and spin density waves in quasi-one-dimensional systems, we develop a method for studying effects of both magnetic and non-magnetic impurities upon these ordered phases. We show that the Anderson-Maki theorem on the insensitivity of the singlet superconducting transition temperature to non-magnetic impurities while showing a marked depression of T_c for magnetic impurities can be generalized to other types of ordered states. We show that the contribution from various scattering terms can be grouped into two classes. One class leaves the mean field transition temperature virtually unchanged while the other causes a marked depression. We discuss the implications of these results on quasi-one-dimensional organic metals.

PREFACE

The intensive research on quasi-one-dimensional conductors, which started in 1973 with the discovery of the compound TTF-TCNQ, has resulted in many theoretical and experimental advances during the past few years, and today the subject has grown into a broad, rich, and respectable field. This development was preceded by a long-standing interest in the problem of ordering in one-dimensional metallic chains, which was motivated by the suggestion of the possibility of superconductivity in such a system, made by

---

H. Gutfreund is Professor of Physics at the Racah Institute of Physics, Hebrew University, Jerusalem. W. A. Little is Professor of Physics at Stanford University.

one of us[1] ten years earlier. The development of this idea was followed by Felix Bloch with keen interest and we both have pleasant recollections of numerous discussions with him on this subject in the middle and late sixties. At that time Felix Bloch was himself working on the Tomonaga model of a one-dimensional Fermi system in an attempt to gain a more basic understanding of superconductivity. In this model one succeeds in treating the full effect of interactions as long as their range is large compared to the Fermi wavelength and their strength small compared to the Fermi energy, while in the BCS theory of superconductivity only the interactions responsible for the pairing correlations are retained. It is interesting to note the full title of Tomonaga's paper:[2] "Remarks on Bloch's Method of Sound Waves Applied to Many-Fermion Problems." Tomonaga refers here to Bloch's paper[3] many years earlier, "Incoherent X-ray Scattering and Density Fluctuations in an Excited Fermi-Gas," published in *Helvetica Physica Acta* in 1934. Indeed, this paper contains and discusses the basic feature of the Tomonaga model, namely, the bosonisation of the Hamiltonian of a one-dimensional Fermi system. All the extensive theoretical work of the last few years on phase transitions in one-dimensional metals is based on this idea.

I. INTRODUCTION

A one-dimensional Fermi system is characterized by a Fermi "surface" consisting of two points at $\pm k_F$. This results in a striking difference between the three- and one-dimensional Fermi systems. While in the first case there exist low-energy electron-hole excitations with momenta between 0 and $2k_F$, in the second case one finds such excitations only in the neighborhood of $q \simeq 0$ and $q \simeq 2k_F$. This means that the important electron-electron interaction processes are those which involve momentum transfers in the neighborhood of these two values. There are four relevant interaction processes, which are represented in figure 1. The $q \simeq 0$ processes may involve either electrons on one side of the Fermi "surface" (g_4) or on both sides (g_2). In both cases the electrons do not change their direction of motion as a result of the interaction between them. For this reason, these processes are referred to as *forward scattering processes*. When electrons on the two sides of the Fermi "surface" exchange momentum $q \simeq 2k_F$, they change their direction of motion. This is therefore a backscattering process (g_1). In the case of one electron per atom (half filled band), there is also an Umklapp process in which two electrons on the same side of the Fermi "surface" scatter together to the other side (g_3). The study of instabilities in a model in which the interaction is reduced to these coupling constants is generally referred to as one-dimensional "g"-ology.

We shall base our discussion on a system described by the electron gas Hamiltonian. By this we assume that the important part of the electron-

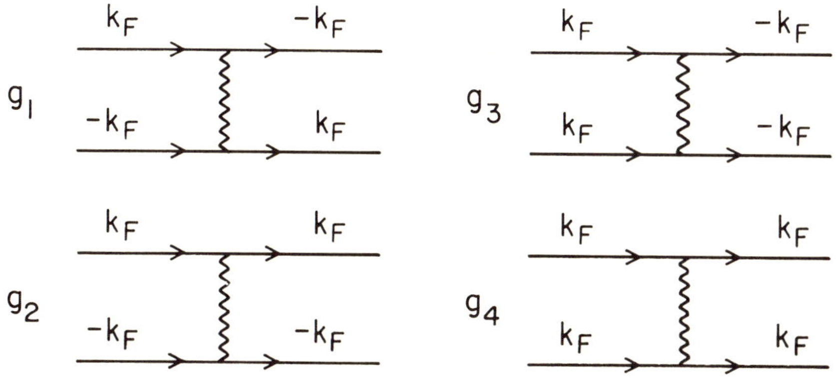

FIG. 1. DIAGRAMMATIC REPRESENTATION OF THE FOUR INTERACTION PROCESSES IN A ONE-DIMENSIONAL FERMI SYSTEM. The forward scattering processes are denoted by g_2 and g_4, the backscattering by g_1 and g_3. The latter exists only for a half-filled band.

phonon interaction is included effectively in the electron-electron interaction. If retardation effects are important, one has to add the electron-phonon interaction explicitly. In one dimension it is convenient to define two electron fields $\psi_{1s}(k) \equiv \psi_s(k_F + k)$ and $\psi_{2s}(k) \equiv \psi_s(-k_F + k)$ for electrons moving in one direction (around $+k_F$) and in the opposite direction (around $-k_F$), respectively, each with spin projection s.

In writing the Hamiltonian we shall break up the interaction into the various processes described above, and neglect the g_3 and g_4 terms. The reason is that g_3 corresponds to a specific case that we omit from our discussion and g_4 serves merely to renormalize the Fermi velocity and has no qualitative effect on the problem of ordering.

Thus, the Hamiltonian describing the one-dimensional system with a linear dispersion relationship is:

$$H - N\mu = \sum\sum \epsilon_k \left(\psi_{1s}^\dagger(k)\,\psi_{1s}(k) - \psi_{2s}^\dagger(k)\,\psi_{2s}(k)\right) + H_{\text{int}}$$

with $\epsilon_k = kv_F$

and $H_{\text{int}} = \frac{1}{2} \sum\limits_{s,s'} \sum\limits_{k,k',p,p'} (g_1 + g_2)$, where

$$g_1 = \left\{\psi_{1s}^\dagger(p')\,\psi_{2s'}^\dagger(p)\,V_1(p',p,k,k')\,\psi_{1s'}(k)\,\psi_{2s}(k')\right\}$$

and

$$g_2 = \left\{\psi_{1s}^\dagger(p')\,\psi_{2s'}(p)\,V_2(p',p,k,k')\,\psi_{2s'}(k)\,\psi_{1s}(k')\right\} \qquad (1)$$

The point of breaking up the interaction terms in this way lies in the fact that the major contributions to the behavior of the system come from small values of k, k', p, and p', and the matrix elements in those neighborhoods may be approximated by the constant terms g_1, g_2, as indicated.

Although a one-dimensional electron system cannot undergo a phase transition at any finite temperature, it may exhibit long range order at $T=0$. An understanding of the conditions for the onset of a particular type of order at $T=0$ and the growth of the corresponding condensate is important for the discussion of the behavior in real filamentary materials.

The main purpose of the present paper is to discuss the effect of impurities on the growth of order in a one-dimensional metal. We shall do that by examining the effect of impurities on the mean-field transition temperature for various types of order in a one-dimensional system. Although the mean-field transition temperature in a strictly one-dimensional system has nothing to do with a phase transition, it does indicate that at this temperature the system begins to develop large fluctuations corresponding to a particular type of order, which increase as $T \to 0$. In a system of coupled one-dimensional chains, the real phase transition temperature is determined by the mean-field T_c and by the nature and magnitude of the interchain coupling.

The onset of order in a system is indicated by nonvanishing "anomalous" averages. In addition to electron-electron "anomalous" averages corresponding to singlet superconductivity (SS) and triplet superconductivity (TS), in a one-dimensional system one also obtains electron-hole "anomalous" averages, corresponding to charge-density waves (CDW) or spin-density waves (SDW). The order (gap) parameters associated with these four types of order are[4]

$$\Delta_{\mathrm{SS}} = -\tfrac{1}{2}(g_1 + g_2) \sum_k \left\langle \psi^\dagger_{1\uparrow}(k)\, \psi^\dagger_{2\downarrow}(-k) - \psi^\dagger_{1\downarrow}(k)\, \psi^\dagger_{2\uparrow}(-k) \right\rangle$$

$$\Delta_{\mathrm{CDW}} = -\tfrac{1}{2}(2g_1 - g_2) \sum_k \left\langle \psi^\dagger_{1\uparrow}(k)\, \psi_{2\uparrow}(k) + \psi^\dagger_{1\downarrow}(k)\, \psi_{2\downarrow}(k) \right\rangle$$

$$\Delta^{0}_{\mathrm{TS}} = \tfrac{1}{2}(g_1 - g_2) \sum_k \left\langle \psi^\dagger_{1\uparrow}(k)\, \psi^\dagger_{2\downarrow}(-k) + \psi^\dagger_{1\downarrow}(k)\, \psi^\dagger_{2\uparrow}(-k) \right\rangle$$

$$\Delta^{\pm 1}_{\mathrm{TS}} = \tfrac{1}{2}(g_1 - g_2) \sum_k \left\langle \psi^\dagger_{1\uparrow}(k)\, \psi^\dagger_{2\uparrow}(-k) \pm \psi^\dagger_{1\downarrow}(k)\, \psi^\dagger_{2\downarrow}(-k) \right\rangle$$

$$\Delta^{0}_{\mathrm{SDW}} = \tfrac{1}{2}\, g_2 \sum_k \left\langle \psi^\dagger_{1\uparrow}(k)\, \psi_{2\uparrow}(k) - \psi^\dagger_{1\downarrow}(k)\, \psi_{2\downarrow}(k) \right\rangle$$

$$\Delta_{SDW}^{\pm 1} = \tfrac{1}{2}\, g_2 \sum \left\langle \psi_{1\uparrow}^{\dagger}(k)\ \psi_{2\downarrow}(k) \ \pm\ \psi_{1\downarrow}^{\dagger}(k)\ \psi_{2\uparrow}(k) \right\rangle \tag{2}$$

In the mean-field treatment of the Hamiltonian[1] each of these gap parameters is different from zero below a certain critical temperature T_c determined by a BCS-type equation. We shall investigate the effect of impurities on this T_c.

The outline of the paper is as follows: In section II we present a general introductory discussion of the effects of impurities. In section III we describe an extension of the matrix formalism of Nambu,[5] which is used in section IV to show that the various scattering mechanisms can be classified into two classes, one which leaves T_c virtually unaffected and one which severely depresses T_c. In section V we present the results of our calculation of the effect of each of the different scattering mechanisms on the various ordered phases. We discuss our results in relation to earlier work. In section VI we point out some of the consequences of our work in relation to ongoing work on quasi-one-dimensional systems.

II. IMPURITIES

It is well known that any disorder caused by the presence of impurities or defects, which are unavoidable in a real material, results in a one-dimensional system in the localization of all the electron states.[6] The most important consequence of this localization is that the static conductivity of such a system vanishes for any localization length.[7] However, in real systems there are several delocalizing effects that allow a non-zero dc conductivity, which may even be of metallic nature. One such mechanism is a finite interchain hopping probability. The delocalizing effect of interchain hopping was recently discussed extensively by Abrikosov and Ryzhkin.[8] Another effect, which may cause coherent electron transport even in a single chain, is the interaction with phonons. This effect was studied by Gogolin et al.[9] Their basic result is that high-frequency phonons with significant dispersion (such that $\omega\tau_{imp} >> 1$ and $\Delta\tau_{imp} >> 1$, where ω is the phonon frequency, Δ is their dispersion, and τ_{imp} is the characteristic time for collisions with impurities) have a delocalizing effect on the electrons. In a range $\tau_{ph} < \tau_{imp}$ (τ_{ph} is the characteristic time for electron-phonon collisions), such phonons increase the localization length and yield a dc conductivity resulting from phonon-assisted diffusion between localized states, whereas at higher temperatures when $\tau_{ph} << \tau_{imp}$, localization is totally destroyed and one obtains the usual Drude picture.

In what follows we shall assume that the Drude regime extends to temperatures below the mean-field T_c for the possible types of order, so that the condensate may build up before the effect of impurities on the normal state properties results in the destruction of the metallic behavior. We shall return to this point in the discussion.

In addition to the effect on the normal properties, impurities in general can have a large effect upon the degree of order that tends to appear as $T \rightarrow 0$ and upon the mean-field transition temperature T_c of any such ordered phases. We shall show that this effect is selective, however, and depending upon the nature of the impurity can severely suppress one type of ordered phase while leaving another virtually unaffected. The result we present is essentially a generalization to all types of order in one-dimensional systems of the Anderson-Maki theorem[10,11] on the insensitivity of the superconducting transition temperature to the presence of non-magnetic impurities. We show that to each type of ordered state in one dimension, whether it be a charge density wave (CDW), spin density wave (SDW), triplet superconductivity (TS), or singlet superconductivity (SS), there exists a type of scattering mechanism that has a negligible effect upon the mean-field temperature of this state. Scattering processes that are not of this type have a severe effect upon T_c. The physical reason for this is as follows: In the pure state each of these ordered phases is characterized by the pairing of two electron states or an electron and a hole state, each of well defined momentum and spin. This pair defines the order parameter of the particular phase. Upon the introduction of the impurity these momentum and spin states will become mixed. Certain types of mixing leave the gap equation, which determines the order parameter of certain of these phases, essentially unchanged. In the case of singlet superconductivity this is illustrated by the minor effect upon T_c of non-magnetic impurities. These non-magnetic impurities mix states of different momenta but leave the gap equation for singlet superconductivity virtually unchanged. This follows because the gap equation results from the pairing of time reversed states and the time reversed nature of the paired states is preserved even in the presence of non-magnetic impurities.

The presence of impurities mixes the plane wave states by contributing terms to the Hamiltonian of the general form

$$V(q) = \int \psi^{\dagger}_{s'}(k+q) \left[\sum_i u_{s's}(x-x_i)\right] \psi_s(k)\ dx$$

$$= \sum_i u_{s's}(q)\, e^{iqx_i} \qquad (3)$$

where $u(x-x_i)$ is the potential of the impurity located at site x_i. The s', s subscripts indicate the effect of the impurity upon the spin.

In one dimension the behavior of the system is dominated by those matrix elements in the vicinity of $q = 0$ and $q = 2k_F$; consequently it is convenient to describe the effects of the impurities in terms of four types of scattering processes. We have considered the following:

(a) forward scattering with no spin flip
(b) forward scattering with spin flip
(c) backward scattering with no spin flip
(d) backward scattering with spin flip.

In general, a given impurity will contribute both forward and back scattering terms, but, depending upon the nature of the impurity, the magnitude of these terms will differ. No real impurity will give forward scattering alone or back scattering alone, but always some mixture of the two. Nevertheless it is convenient to separate the contributions of the scattering into these two components in treating the effect of impurities on the ordered states.

III. "TRIPLE" NAMBU FORMALISM

In order to take into account SDW, CDW, SS, and TS together with both magnetic and non-magnetic impurity scattering, it is convenient to use an extension of Nambu's matrix representation.[5] We define an 8-component spinor field

$$\Psi(k) \equiv \Big(\psi^+_{1\uparrow}(k), \psi^+_{1\downarrow}(k), \psi^+_{2\uparrow}(k), \psi^+_{2\downarrow}(k), \psi_{1\uparrow}(-k), \psi_{1\downarrow}(-k), \psi_{2\uparrow}(-k), \psi_{2\downarrow}(-k)\Big) . \qquad (4)$$

The Hamiltonian[1] may be written in a compact form in terms of 8×8 matrices acting upon the above 8-component spinor in a similar way to that done by Horovitz[4] with 4-component spinors used to describe the phase transitions alone. The inclusion of magnetic impurity scattering requires the extension to the 8-component description in our case. Likewise, the "anomalous" averages[2] and the impurity scattering interactions may be expressed by these matrices. The 8×8 matrix in this "triple" Nambu representation is generated by the direct product of three Pauli or unit matrices $\sigma_i \cdot \tau_j \cdot \varrho_k$, where σ describes spin changes, τ scattering across or at the Fermi surface, and ϱ particle-hole changes. The order of the terms of the direct product matrix used here is defined as:

$$\sigma_1 \,.\, \tau_3 \,.\, \varrho_1 \;\equiv\; \left(\begin{array}{cc|cc} & & \sigma_1 & \\ & & & -\sigma_1 \\ \hline \sigma_1 & & & \\ & -\sigma_1 & & \end{array}\right) \tag{5}$$

The presence of the anomalous averages contributes terms to the single particle Green's function, which by analogy to Nambu's method may be described in terms of an eight-by-eight matrix of thc form

$$\tilde{G}^{-1}(p) \;=\; i\omega(1.1.1) - \epsilon_p(1.\tau_3.1) - \tilde{\Sigma}(p) \tag{6}$$

The self energy $\tilde{\Sigma}(p)$ may be expanded in terms of the 64 matrices, of the general form $\tilde{\Gamma}_{ijk} \equiv \tilde{\sigma}_i \,.\, \tilde{\tau}_j \,.\, \tilde{\varrho}_k$ to give

$$\tilde{\Sigma}(p) \;=\; \sum_{i,j,k} \Delta_{ijk}\, \tilde{\Gamma}_{ijk} \tag{7}$$

where the coefficients Δ_{ijk} are proportional to averages of bilinear expressions in the original fields. We shall not interpret all of these here, but will concentrate on the anomalous averages that correspond to the various order parameters.

The matrices associated with these various coefficients are found to be the following:

$$\begin{aligned}
\Gamma_{SS} &= i\sigma_2 \,.\, \tau_1 \,.\, \varrho^+ & \qquad {}^0\Gamma^u_{SDW} &= \sigma_3 \,.\, i\tau_2 \,.\, 1 \\
{}^0\Gamma_{TS} &= \sigma_1 \,.\, i\tau_2 \,.\, \varrho^+ & \qquad {}^0\Gamma^g_{SDW} &= \sigma_3 \,.\, \tau_1 \,.\, \varrho_3 \\
{}^1\Gamma_{TS} &= 1 \,.\, i\tau_2 \,.\, \varrho^+ & \qquad {}^1\Gamma^u_{SDW} &= \sigma_1 \,.\, i\tau_2 \,.\, 1 \\
{}^{-1}\Gamma_{TS} &= \sigma_3 \,.\, i\tau_2 \,.\, \varrho^+ & \qquad {}^1\Gamma^g_{SDW} &= \sigma_1 \,.\, \tau_1 \,.\, \varrho_3 \\
\Gamma^u_{CDW} &= 1 \,.\, i\tau_2 \,.\, 1 & \qquad {}^{-1}\Gamma^u_{SDW} &= i\sigma_2 \,.\, \tau_1 \,.\, 1 \\
\Gamma^g_{CDW} &= 1 \,.\, \tau_1 \,.\, \varrho_3 & \qquad {}^{-1}\Gamma^g_{SDW} &= i\sigma_2 \,.\, i\tau_2 \,.\, \varrho_3
\end{aligned} \tag{8}$$

For the CDW and SDW we have constructed odd and even combinations (u, g) of the order parameters defined in (2) with their respective hermitian conjugates. The u and g states in general are affected differently by impurities. For the superconducting order parameters, the odd and even combinations have the same behavior, and hence such a distinction is unnecessary.

Finally, we express in the matrix representation the contribution to the Hamiltonian from the four types of impurity interaction enumerated in the preceding section.

$$\text{(a)} \quad \sum_{i=1,2} \sum_{s} \sum_{k,q} \psi^{\dagger}_{is}(k+q)\, V_1(q)\, \psi_{is}(k)$$

$$= \tfrac{1}{2} V_1(0) \sum_{k,q>0} \Psi^{\dagger}(k+q)\,(1.1.\varrho_3)\,\Psi(k)$$

$$\text{(b)} \quad \sum_{i=1,2} \sum_{s} \sum_{k,q} \psi_{is}(k+q)\, V_2(q)\, \psi_{i,-s}(k)$$

$$= \tfrac{1}{2} V_2(0) \sum_{k,q>0} \Psi^{\dagger}(k+q)\,(\sigma_1.1.\varrho_3)\,\Psi(k)$$

$$\text{(c)} \quad \sum_{s} \sum_{k,q} \Big[\psi^{\dagger}_{1,s}(k+q)\;\; V_3(2k_F+q)\;\; \psi_{2,s}(k) \;+\; \psi^{\dagger}_{2,s}(k+q)\;\; V(2k_F)\;\; \psi_{1,s}(k)\Big]$$

$$= \tfrac{1}{2}\;\; V_3(2k_F) \sum_{k,q>0} \Psi^{\dagger}(k+q)\;\; (1.i\tau_2.\varrho_3)\;\; \Psi(k)$$

$$\text{(d)} \quad \sum_{s} \sum_{k,q} \Big[\psi^{\dagger}_{1,s}(k+q)\, V_4(2k_F+q)\, \psi_{2,-s}(k) + \psi^{\dagger}_{2s}(k+q)\, V_4(2k_F+q)\, \psi_{1-s}(k)\Big]$$

$$= \tfrac{1}{2} V_4(2k_F) \sum_{k,q>0} \Psi^{\dagger}(k+q)\,(\sigma_1.\tau_1.\varrho_3)\,\Psi(k) \qquad (9)$$

We treat now the effects of these scattering terms on the ordered states using perturbation theory.

IV. EFFECTS UPON T_c

In this section we follow closely the method of treatment of impurity scattering on an ordered phase described by Maki.[11] In general the temperature Green's function for the electrons in the presence of the impurities has the form

$$G^{-1}(p, \omega_n) = i\tilde{\omega}_n(1.1.1) - \tilde{\epsilon}_p(1.\tau_3.1) - \tilde{\Delta}_\nu\Gamma_\nu \tag{10}$$

where Γ_ν is the matrix corresponding to the order parameter Δ_ν existing in the system. Here $\tilde{\omega}_n$, $\tilde{\epsilon}_n$, and $\tilde{\Delta}_\nu$ are the renormalized frequencies, energies, and order parameters, respectively, determined by the equation.

$$G^{-1}(p,\omega_n) = \left[G^0(p,\omega_n)\right]^{-1} - \Sigma(p,\omega_n) \tag{11}$$

where $$\left[G^0(p,\omega_n)\right] = i\omega_n(1.1.1) - \epsilon_p(1.\tau_3.1) - \Delta_\nu\Gamma_\nu \tag{12}$$

and $\omega_n = 2\pi T(n + ½)$.

The self energy, $\Sigma(p,\omega_n)$, has contributions from terms of various powers in the scattering term $V(q)$. The linear term causes a shift of the individual energy levels. We transform to a new representation and new Fermi energy such that the energy, $\tilde{\epsilon}_p$, in the presence of these impurities is given by $\tilde{\epsilon}_p(1.\tau_3.1)$. This will result in a change of the density of states whose specific form depends upon the particular distribution of impurities. The next term is the quadratic term, which we retain. To this approximation the self energy $\Sigma(p,\omega_n)$ is given by

$$\Sigma(p,\omega_n) = \frac{1}{2\pi}\int V(p-p')\,G(p',\omega_n)V(p-p')\,dp' \tag{13}$$

where $V(p-p') = V_i(p-p')\,M$ where V_i is one of the terms (a), (b), (c), or (d) in (9) and M is the corresponding matrix in each of these equations. We use the standard convention describing the self energy by $\Sigma(p,\omega_n)$ and trust that the context of the equations will distinguish it from the summation symbol $\sum$. We note that in all cases $M^2 = 1$.

Using (10), we rationalize $G(p)$ in (13) assuming Δ_ν is a constant and obtain

$$\underset{\sim}{G}(p,\omega_n) = -\frac{i\tilde{\omega}_n(1.1.1) + \tilde{\epsilon}_p(1.\tau_3.1) + \tilde{\Delta}_\nu\Gamma_\nu}{\tilde{\omega}_n^2 + \tilde{\epsilon}_p^2 + \tilde{\Delta}_\nu{}^2} \tag{14}$$

where we utilize the fact that $(1.\tau_3.1)$ anti-commutes with every Γ_ν defined in Eq. (8). Assuming $V_i(p-p')$ is slowly varying in the vicinity of $p' = k_F$, we can integrate over p', and using (14) in (13) we obtain

$$\Sigma(p,\omega_n) = -\frac{1}{2\tau_i}\frac{i\tilde{\omega}_n(1.1.1) \pm \tilde{\Delta}_\nu\Gamma_\nu}{\sqrt{\tilde{\omega}_n^2 + \tilde{\Delta}_\nu^2}}, \tag{15}$$

where the + (−) sign depends upon whether M anti-commutes (commutes) with Γ_ν and τ_i is the scattering time given by

$$\frac{1}{\tau_i} = \frac{1}{v_F} V_i^2 . \tag{16}$$

Substituting (15) in (11) we get

$$\tilde{\omega}_n = \omega_n + \frac{1}{2\tau_i}\frac{\tilde{\omega}_n}{\sqrt{\tilde{\omega}_n^2 + \tilde{\Delta}_\nu^2}} \tag{17}$$

$$\tilde{\Delta}_\nu = \Delta_\nu \pm \frac{1}{2\tau_i}\frac{\tilde{\Delta}_\nu}{\sqrt{\tilde{\omega}_n^2 + \tilde{\Delta}_\nu^2}} \tag{18}$$

$$\tilde{\epsilon}_p = \epsilon_p + (\text{corrections due to terms linear in } V). \tag{19}$$

The gap equation at finite temperatures is

$$\Delta = \frac{gT}{v_F}\sum_{\omega_n}^{\infty}\int_{-\epsilon_c}^{\epsilon_c}\frac{\tilde{\Delta}\, d\epsilon}{\tilde{\omega}_n^2 + \epsilon^2 + \tilde{\Delta}^2} \tag{20}$$

where g, the net interaction, is the particular combination of g_1, g_2, etc., responsible for the particular type of order Δ, and ϵ_c is the energy cut-off in this interaction. One may verify that the integral over the energy may be extended to infinity if at the same time the sum over ω_n is cut off at a value of $n_m = \dfrac{\epsilon_c}{2\pi T}$. Carrying out the energy integral then we get

$$\Delta = \frac{2\pi gT}{v_F}\sum_{n=0}^{n_m}\frac{\tilde{\Delta}}{\sqrt{\tilde{\omega}_n^2 + \tilde{\Delta}^2}} . \tag{21}$$

Following Maki we set $\tilde{\omega}_n/\tilde{\Delta} = u_n$ and (21) becomes

$$\Delta = \frac{2\pi g T}{v_F} \sum_{n=0}^{n_m} \frac{1}{\sqrt{1 + u_n^2}} . \tag{22}$$

Near T_c the gap is small and u_n is large. An expansion can be made for u_n in powers of Δ. Using (17) and (18) we can express ω_n/Δ in terms of $\tilde{\omega}_n/\tilde{\Delta}$. We find that if *M anti-commutes* with Γ_ν (+ sign in (18)) then

$$\frac{\omega_n}{\Delta} = \frac{\tilde{\omega}_n}{\tilde{\Delta}} = u_n \tag{23}$$

and if *M commutes* with Γ_ν

$$\frac{\omega_n}{\Delta} = u_n \left[1 - \frac{1}{\tau_i \Delta \sqrt{1 + u_n^2}} \right] . \tag{24}$$

If M for a particular type of scattering from an impurity anti-commutes with Γ_ν, then (22), because of (23), gives the same critical temperature and gap equation except for the effects due to changes in the density of states of $\tilde{\epsilon}_p$ relative to that of ϵ_p. On the other hand, if M commutes with Γ_ν, then (24) in general gives a reduced T_c. This latter point can be seen from the following: We note that u_n will be large near T_c and hence $(1 + u_n^2)^{-1/2}$ can be expanded in powers of $1/u_n$. Using (24) and solving for u_n we find:

$$\frac{1}{\sqrt{1 + u_n^2}} = \frac{\Delta}{\omega_n + \alpha} - \tfrac{1}{2} \times \left[\frac{1}{(\omega + \alpha)^3} - \frac{\alpha}{(\omega + \alpha)^4} \right] \Delta^3 + \ldots \tag{25}$$

where $\alpha = 1/\tau_i$.

At T_c we have

$$1 = \frac{2\pi g\ T_c}{v_F} \sum_{n=0}^{n_m} \left(\frac{1}{\omega_n + \alpha} \right) = \frac{g}{v_F} \sum_{n=0}^{n_m} \left(\frac{1}{n + \tfrac{1}{2} + \varrho_c} \right) \tag{26}$$

with $\omega_n = 2\pi T(n + \tfrac{1}{2})$

$$\varrho_c = \alpha/2\pi T_c = \frac{1}{2\pi\tau_i T_c} \text{ and } n_m = \frac{\epsilon_c}{2\pi\ T_c} .$$

In the absence of impurities $\varrho_c = 0$ and

$$1 = \frac{2\pi g\ T_{co}}{v_F} \sum_{n=o}^{n_m'} \frac{1}{\omega_n} = \frac{g}{v_F} \sum_{o}^{n_m'} \left(\frac{1}{n + 1/2}\right)$$

where $n_m' = \dfrac{\epsilon_c}{2\pi\ T_{co}}$. (27)

In terms of the digamma function $\psi(z)$[12]

$$\psi(z) = \mathrm{Lt}_{n\to\infty} \left[\log n - \sum_{o}^{n} \frac{1}{m + z}\right] . \quad (28)$$

(26) and (27) give us approximately

$$\log \left(n_m'/n_m\right) = \psi(1/2) - \psi(1/2 + \varrho_c) . \quad (29)$$

Hence

$$\log \left(T_c/T_{co}\right) = \psi(1/2) - \psi(1/2 + \varrho_c) \quad (30)$$

$$T_c = T_{co} \exp - \left\{\psi(1/2 + \varrho_c) - \psi(1/2)\right\} . \quad (31)$$

For small values of ϱ_c equation (31) gives a linear depression of T_c with $1/\tau_i$

$$T_{co} - T_c = \frac{\pi}{4\tau_i} . \quad (32)$$

From this we note that the *relative* depression of T_c with impurity concentration

$$\frac{T_{co} - T_c}{T_{co}} = \frac{\pi}{4\tau_i\ T_{co}} \quad (33)$$

is inversely proportional to T_{co} so that phases with high transition temperatures are less affected by impurity scattering than those of low transition temperature.

Noting that ϱ_c itself depends upon T_c, one finds that the transition temperature is suppressed to 0 K for values of $\alpha = \tau_i^{-1}$ such that $\alpha \geq \dfrac{\pi\ T_{co}}{2\gamma}$ where $\gamma = e^C = 1.78$; $C = 0.5772$, Euler's Constant.

V. EFFECTS OF IMPURITY SCATTERING ON THE ORDERED PHASES

From the foregoing we see that the mean-field transition temperature of a given phase will be depressed according to Eq. (31) by the presence of impurities if the matrix that describes the type of scattering mechanism (Eq. (9)) commutes with the matrix for the ordered phase (Eq. (8)). On the other hand, if these matrices anti-commute, then T_c remains virtually unchanged. In table I we list the commutation properties of these pairs of matrices. The set describing forward scattering without spin flip warrants further comment. Zavadovski[13] had shown that by treating forward scattering by impurities in the quasi-classical approximation, neither singlet superconductivity nor the charge-density-wave is affected by such impurities. This appears to be at variance with the results of table I. The explanation of this difference is as follows: If we include in our initial Hamiltonian the forward scattering terms, then we can obtain the one-electron states and energies in the presence of these terms by using the WKB approximation.[14] The plane wave states become phase modulated,

$$\psi_k(x) = \exp i \left[kx - \frac{1}{2k} \int^x u(x)dx \right] . \tag{34}$$

For the states in the immediate vicinity of the Fermi surface the phase factors will be all the same. We may now use these phase modulated states instead of the plane wave states as the basis of our representation (4). Taking the anomalous averages into account, one can calculate the mean-field transition temperature for condensation into the various paired states. The resultant gap equation and values for T_c will be essentially the same as for the plane wave states. This is the essence of the Zavadovski result. On the other hand, in this paper we have considered a specific form of pairing described by (4) formed from *plane wave states* k and $-k$. The transition temperature for condensation into such plane wave states is indeed suppressed and the plane wave condensate is replaced by that of the phase-modulated states. Thus the two different results are in fact in agreement with one another.

The results of our analysis of the effects of forward scattering with no spin flip we place in braces in table I to indicate that the T_c of the original *plane wave* ordered states will indeed be depressed for those cases where the scattering terms commute. Condensation will not occur into these, however, since the transition temperature of the phase modulated ordered states will be higher, and T_c for these will be essentially the same as in the absence of the impurities. Thus the effects of forward scattering can be transformed away.

TABLE 1

Ordered State / Scattering Mechanism	SS	CDW		TS			SDW					
	0	0		0	+1	−1	0		+1		−1	
		even	odd				even	odd	even	odd	even	odd
Forward No Spin Flip	(A)	(C)	(C)	(A)	(A)	(A)	(C)	(C)	(C)	(C)	(C)	(C)
Forward Spin Flip	C	C	C	A	A	C	A	A	C	C	A	A
Backward No Spin Flip	A	C	A	C	C	C	C	A	C	A	C	A
Backward Spin Flip	C	C	A	C	C	A	A	C	C	A	C	A

The argument above shows up a weakness in the approach we have taken, for we are forced to make an assumption through our initial choice of representation (4) as to the nature of the condensed states and then to consider the effects of scattering upon these. A more powerful approach to the problem would be to determine the one-electron states in the presence of the impurities *first*, and then to consider possible superconducting or spin and charge density wave condensates formed from these states. This, as we have shown, can be done to a good approximation for forward scattering but is much more difficult to do for other types of scattering.

IV. DISCUSSION

Our formalism provides a compact way to see the effects of impurities upon the various order parameters. For singlet superconductivity we see that impurities which cause both forward and backward scattering, but without spin flip, leave T_c for this phase essentially unchanged. Spin flip scattering severely suppresses T_c, however, as one would expect from the known behavior in three-dimensional metals. These results stress the importance of eliminating paramagnetic impurities, dangling bonds, and unpaired spins, which might interact with the conduction electrons in such materials as $(CH\ I_{0.22})_x$, HMTSF − TCNQ, or other quasi-one-dimensional organic materials that are metallic at low temperature but that might become superconducting instead through the elimination of such scattering terms.

We note that chemical impurities that cause back scattering have a disastrous effect upon triplet state superconductivity. Thus polymer cross linking or other discontinuities or impurities would strongly suppress superconductivity of this type. Because of this, in those regions of the g-ology diagram[15] where TS and SS can co-exist, the presence of such scattering would cause the SS region to encroach upon the TS region. Scattering with spin flip affects the different TS components differently, but as any magnetic scattering center is likely to cause both forward and backward scattering with both spin flip and non-spin flip contributions, the net effect of such impurities would be to suppress all the TS components.

The various odd and even components of the CDW and SDW are affected by impurities in what appears to be a rather complicated manner. The underlying physics, however, is relatively simple. To first order the scattering terms in the Hamiltonian introduce into the plane-wave states a small admixture of other such states. When one constructs the bilinear products of these to form the condensates, the admixtures can cancel or add depending on whether the particular condensate combination is odd or even (u or g). If the admixture to the condensate is of a different symmetry to that of the condensate itself *and* the contributions do not cancel, then T_c is depressed. Otherwise T_c remains virtually unaffected.

We wish to remind the reader that we have considered here only the effect of impurities on the growth of order in a single chain. Phase transitions in filamentary systems occur when the phases of the order parameters on different chains become correlated through interchain coupling, and long range order develops in the transverse direction. Impurities have an important effect on the build-up of order in the transverse direction,[16] and this has to be taken into account in the discussion of phase transitions in real quasi-one-dimensional materials.

Finally, we recall that the whole discussion is based on the assumption that the electrons are delocalized. We wish to point out that, at least in molecular crystals, this may occur already at rather low temperatures. It follows from the work of Gogolin et al.[9] that in the presence of weak phonon scattering, namely when $\tau_{ph}^{-1} < \tau_{imp}^{-1}$, the hopping conductivity is of the form $\sigma \propto \tau_{imp}^{2}/\tau_{ph}$, where $\tau_{ph}^{-1} = \tau_1^{-1} + \tau_2^{-1}$; τ_1 and τ_2 are the relaxation times corresponding to backward and forward phonon scattering respectively. Thus, forward scattering phonons are as effective as backscattering phonons in causing delocalization. When $\tau_{ph}^{-1} > \tau_{imp}^{-1}$ one gets the usual Drude behavior with $\sigma \propto (\tau_{imp}^{-1} + \tau_1^{-1})^{-1}$. In this region τ_2^{-1} does not contribute to the resistivity. In the first region, when $\tau_{ph}^{-1} < \tau_{imp}^{-1}$, the conductivity increases with T until it reaches a maximum. If $\tau_2^{-1} >> \tau_1^{-1}$, this maximum is reached at much lower temperatures than one would estimate from the relaxation time derived from the conductivity in the metallic region. It was shown by Kaveh et al.[17] that this is the case for scattering of electrons by molecular vibrations. This follows merely from the fact that a small Fourier component (forward scattering) of an extended potential is much larger than the $2k_F$ Fourier component (back scattering). This observation was used by the authors of ref. 17 to explain the metallic behavior of the organic metals like TTF-TCNQ.

ACKNOWLEDGMENT

This work was supported by a grant from the National Science Foundation (Grant No. DMR 74-00427-A03), the National Aeronautical and Space Agency (Contract No. JPL 953752), and U.S.-Israeli Binational Science Foundation (Grant No. 215).

REFERENCES

1. W. A. Little, Phys. Rev. **134A**, 1416 (1964).
2. S. Tomonaga, Prog. Theor. Phys. **5**, 544 (1950).
3. F. Bloch, Helv. Phys. Acta **7**, 385 (1934).

4. B. Horovitz, Solid State Commun. **18**, 445 (1976).
5. Y. Nambu, Phys. Rev. **117**, 648 (1960).
6. N. F. Mott and W. D. Twose, Adv. Phys. **10**, 107 (1961).
7. B. I. Halperin, Adv. Chem. Phys. **13**, 123 (1967).
8. A. A. Abrikosov and I. A. Ryzhkin, Adv. Phys. **27**, 147 (1978).
9. A. A. Gogolin, V. I. Mel'nikov and E. I. Rashbra, Sov. Phys.—JETP **42**, 168 (1976) (Zh. Eksp. Teor. Fiz. **69**, 327 (1975)).
10. P. W. Anderson, J. Phys. Chem. Solids **11**, 26 (1959).
11. K. Maki, in *Superconductivity*, vol. 2, p. 1035, ed R. D. Parks (Marcel Dekker, New York, 1969).
12. H. Jeffreys and B. Jeffreys, *Methods of Mathematical Physics* (Cambridge University Press, Cambridge, 1972), pp. 465-466. Note that the convention used here is that of Maki where the digamma function $\psi(z+1)$ is that denoted by $F(z)$ by Jeffreys and $\Psi(z)$ by E. Jahnke and F. Emde, *Tables of Functions* (Dover Publications, New York, 1954), pp. 18-20.
13. A. Zavadovski, Sov. Phys.—JETP **27**, 767 (1968).
14. P. M. Morse and H. Feshbach, *Methods of Theoretical Physics* (McGraw Hill Book Co., New York, 1953), pp. 1092-1095.
15. H. Gutfreund and R. A. Klemm, Phys. Rev. **B14**, 1073 (1976).
16. A. I. Larkin and V. I. Mel'nikov, in *Organic Conductors and Semiconductors*, Lecture Notes in Physics 65 (Springer Verlag, Berlin, 1977), p. 177.
17. M. Kaveh, M. Weger, and H. Gutfreund, Solid State Commun., in press.

HERITAGE OF THE BLOCH EQUATIONS IN QUANTUM OPTICS

by E. L. Hahn

ABSTRACT

The Bloch equations as originally applied to nuclear magnetic resonance have wide application to a variety of physical effects that do not necessarily involve gyromagnetic spin systems. In particular, they are useful in predicting effects involving electric dipole laser resonance phenomena, which in many cases are analogs of effects observed in nuclear magnetic resonance.

I. INTRODUCTION

Among the versatile contributions of Felix Bloch to physics, the Bloch equations of nuclear induction[1] (phrased more commonly today as "nuclear magnetic resonance" [NMR]) have played a major role in guiding researchers and students in the interpretation of resonance experiments. The direction of my research in physics would have been quite different without recourse to the Bloch equations. As a graduate and one-year post-doctoral student at the University of Illinois, my start in NMR research was based on these equations. Since I did not know how to use quantum mechanics, the classical pictures allowed by Bloch's equations saved my neck because they afforded a reasonable macroscopic interpretation of spin echo experiments. It was my good fortune to continue these experiments at Stanford, and there my association with Felix had a profound effect on the way I have done physics ever since. At the beginning of my two years (1950–1952) at Stanford, I remember distinctly that my propensity for arm-waving vector models somewhat overwhelmed Felix. Nevertheless, with dis-

E. L. Hahn is Professor of Physics at the University of California, Berkeley.

cussions seasoned every now and then with dubious jokes, we soon developed a rapport. Eventually, I learned how to use quantum mechanics from Felix after numerous priceless sessions at the blackboard. His effectiveness as a teacher was evident from his broad ability to shift back and forth between the thinking of phenomenological models and fancy formalism.

The Bloch equations have wide application to a number of physical effects that do not necessarily involve gyromagnetic spins. They apply directly to any two quantum level or equally spaced quantum level system, and are particularly useful in predicting effects involving electric dipole laser resonance phenomena. It follows also that the Bloch-type equations are generally descriptive of systems that contain bistable states. For example, the phase transition in ferroelectric KH_2PO_4, associated with the two potential well model of the hydrogen bond, can be deduced[2] from a set of Bloch equations. Free electron laser action from a relativistic electron beam that passes through a spatially periodic magnetic field can be analyzed[3] in terms of coupled Bloch-Maxwell equations. In special situations concerning resonance radiation interacting with unequally spaced multilevel systems, particularly three level ones (see figures 6 and 7, pp. 36-37), such systems can be formulated in terms of the density matrix to give an equivalent two level Bloch equation description.[4] Of course, it is common knowledge that the Bloch equations have a historical root in the mechanical torque equation for the motion of a physical top. The torque equation for the description of gyromagnetic resonance phenomena as presented by Bloch provides a powerful phenomenological means for connecting the linear and nonlinear regimes of resonance response in one formulation. His equations introduced the idea that the otherwise coherent time dependence of quantum mechanical macroscopic moment expectation values could be coupled to phenomenological damping terms. They describe resonance and dispersion in classical terms that connect with the quantum mechanical interpretation through the Liouville equation for the density matrix from which they are derived. Bloch's famous damping time constants T_1 (longitudinal spin lattice relaxation time) and T_2 (transverse phase memory total relaxation time) are common physics language of today among resonance researchers. Although the original phenomenological Bloch equations work very well for fluids, in many cases they are not rigorous for all systems, particularly solids. Nevertheless, the Bloch formulation has stimulated new statistical investigations with the density matrix that are more rigorous for the particular system under investigation. Exceedingly useful is the property that the Bloch equations enable predictions of nonlinear quantum macroscopic phenomena[5] that no amount of fastidious quantum mechanical perturbation theory could predict as handily.

II. THE HERITAGE OF SPIN RESONANCE IN QUANTUM OPTICS

Numerous effects in quantum optics known today are analogs of spin resonance phenomena discovered in former days and are often predicted by Bloch equations. Although considerations of propagation and fluorescence are not present in NMR cavity resonance experiments, there remains a host of similar effects in the time and frequency domain that also appear in quantum optics. One can list as examples adiabatic inversion, Rabi (nutation) oscillations, Bloch-Siegert shift, free precession echo effects, and superposition states. Researchers in quantum optics are sometimes unaware that they rediscover old NMR phenomena. Or they may have a tendency to ignore similar known NMR effects as unimportant because quantum optical effects are governed by matrix elements anywhere from 10^8 to 10^{15} times larger than those in NMR. I am reminded of the story of a visit by the Sultan of Morocco to Berkeley. After he was introduced to a number of high energy physicists, he chided another man, who said he was a low energy physicist, for being too modest! It is my intent in this essay to discuss with immodesty a few points about the impact of low energy magnetic resonance on the development of quantum optics.

Coherent quantum spectroscopy was initiated by the Rabi[6] atomic molecular beam radio frequency (rf) resonance method. For the first time coherent electromagnetic radiation was applied to excite atomic spin resonance transitions. A coherent spectroscopy was achieved with unprecedented resolution, and yet it was apparatus-limited because of the finite beam excitation time in its path from source to detector. (Later a modification by Ramsey[7] reduced this effect considerably.) The Bloch[1] and Purcell[8] groups introduced the science of NMR in condensed matter by extending the Rabi resonance principle to the continuous excitation of spin ensembles in a resonant cavity. The higher resolution capability of this approach very soon revealed sharp NMR spectra; and numerous effects on the resonance by the chemical and solid state environment were discovered. Microwave electric dipole spectroscopy of rotational and vibrational states in molecules showed a parallel development. During these early investigations, the discovery of new spectra, spins, moments, line shifts, and line shape characteristics dominated the attention of resonance research workers. Interest was focused on atomic properties and not on changes in the resonance radiation itself, which exchanges energy with the excited atoms. With the applied radiation fields assumed as fixed driving terms in the analysis, a number of physical ideas that might have led to an earlier realization of the maser-laser principle were suppressed. The impasse was broken somewhat when the idea of coherent radiation reaction, though quite old, was given prominence by Dicke[9] (1954) with regard to coherent emission of electric dipole radiation. He championed the point that an ensemble of atomic electric dipole moments radiates coherently in the same manner as a precessing spin ensemble, and that coherently excited electric dipoles can be identically described by the same superposition coherence states that apply to NMR. It was nothing new, and yet it was instructive to reiterate that any resonance signal in the presence of a driving field (stimulated absorption or emission), or a coherently emitted field from an array of phased dipoles in the absence of a driving field, is the result of a reaction field produced by a macroscopic radiating source polarization term in Maxwell's equations. In addition, Dicke posed a very interesting new question: How does an ensemble of atomic particles in the completely inverted excited state evolve from a condition of incoherent spontaneous emission to a condition of coherent spontaneous emission, where only the latter emission is predicted by Maxwell's equations and the former emission is treated by quantum electrodynamics? Unfortunately, while Dicke meant the term "superradiance" to be applied to this particular evolution process from incoherence to coherence, "superradiance" is now applied to any process of coherent emission involving light. Of course by use of the Bloch equations coupled to a tuned circuit, superradiance would then apply as well to NMR free precession effects.[10] In NMR an oscillating reaction field is produced by a current, which is induced in an inductance by a precessing magnetization according to Faraday's law. For ordinary NMR experiments the effect of the reaction field (see figure 5, p. 35) in directly altering the Zeeman energy of the coherent precessing magnetization is imperceptible (except under very special circumstances[11]) because of the dominating effects of shorter damping times T_1, T_2 and of inhomogeneous broadening. Yet it is this very change in energy that makes possible the observation of NMR signals, whether they result from driven resonance or from free precession. In quantum optics, rapidly varying electric reaction fields from dense ensembles of electric dipole moments can be sufficiently intense to "tip" the polarization completely to the ground state (or to the excited state) in times short compared to the optical T_1 and T_2 time constants. In an inverted two level system, the electric fields that derive from stored energy in the atoms of sufficient density are intense enough to act like "180° pulses," which convert populations completely from one state to the other. This is the mechanism inherent in the self-induced transparency phenomenon.[12]

Although Dicke's paper contained all the ingredients that showed the equivalence of an ensemble of spin $I = ½$ particles to that of an optical two level system, his analysis in using the rules of angular momentum did not explicitly spell out the torque equation of motion for a giant macroscopic electric dipole moment. In 1957, Feynman, Vernon, and Hellwarth[13] showed explicitly that the Bloch equation form was applicable to a two level system with any type of dipole matrix element, involving circularly polarized radiation resonance (angular momentum change $\Delta m = \pm 1$) or linearly polarized radiation resonance ($\Delta m = 0$). Although the discussion in their paper was restricted to a particular MASER problem, they promoted the viewpoint of the semi-classical method, in which the electric field is a self-consistent solution of both the optical type Bloch equations, originating from the density matrix, and Maxwell equations. The work of Jaynes[14] and Lamb[15] extended this semi-classical approach later to quantum maser and laser analysis. And to complete this brief history, the maser-laser principle was finally realized in 1954 (Townes,[16] Basov and Prokhorov,[17]), when proposals for pumping to maintain inverted quantum states combined with efficient coherent radiation feedback were implemented in the microwave ammonia maser and later in an optical solid state system.

III. THE OPTICAL BLOCH-MAXWELL EQUATIONS

The Bloch-Maxwell semi-classical equations of quantum optics consist of a set of equations in the form of Bloch's original nuclear induction equations, which describe the dynamics of optical electric polarization of a two level system, coupled to Maxwell's equations. In addition to the usual time dependence, the spatial dependence of the electric polarization and the propagating electric field must be considered over two level medium dimensions greater than the optical wavelength $\lambda = 2\pi/k$, where $k = \omega/c$ is the propagation constant in vacuum, and ω is the optical frequency. The original Bloch equations

$$\frac{d\mathbf{M}}{dt} = \gamma(\mathbf{M} \times \mathbf{H}) - \frac{\hat{x}_o M_x + \hat{y}_o M_y}{T_2} - \frac{\hat{z}_o(M_z - M_o)}{T_1} \qquad (1)$$

for the macroscopic magnetization **M** apply to an ensemble of N spins/cc subjected to a magnetic field

$$\mathbf{H} = \hat{z}_o H_o + \hat{x}_o H_1(t).$$

Define the applied linear polarized rf field along the $\hat{x}_o$ direction as

$$H_1(t) = 2H_1 \cos \omega t$$

perpendicular to the polarizing field $\hat{z}_o H_o$. As usual we specify γ as the gyromagnetic ratio, $\mu_o = \gamma\hbar I$ the magnetic moment, I the spin quantum number, and $\omega_o = \gamma H_o$ as the transition Larmor frequency. Angular momentum changes $\Delta m = \pm 1$ correspond to the choice of either sign of the gyromagnetic ratio $\pm |\gamma|$. A particular sign signifies the secular interaction of right (+) or left (−) circularly polarized fields. A sum of opposite senses of precessing polarizations describes the dynamics of a system undergoing $\Delta m = 0$ transitions.

The Hamiltonian translation of terms in Eq. (1) to electric dipole notation is presented as follows: For the total Hamiltonian let

$$\mathcal{H} = -\gamma\hbar I_z H_o - \gamma\hbar I_x H(t) \rightarrow \mathcal{H}_o - 2\sigma_x p_o \mathcal{E} \cos\left[\omega t - kz + \phi(z,t)\right],$$

where $\phi(z,t)$ is an arbitrary propagation phase assigned to a plane wave propagating in the z direction, and $\boldsymbol{\sigma}$ is the Pauli operator. Important equivalences are

$$\mu_o \rightarrow p_o,$$

$$\gamma \rightarrow \varkappa = 2p_o/\hbar,$$

$$\gamma\hbar I_x \rightarrow p_o\sigma_x = (p_o/2)(\sigma_+ + \sigma_-),$$

$$\mathbf{M} \rightarrow \mathbf{P},$$

$$H_1(t) \rightarrow 2\mathcal{E} \cos[\omega t - kz + \phi(z,t)],$$

$$\text{Energy: } W = -M_z H_o \rightarrow N\,\mathrm{Tr}\{\mathcal{H}_o\varrho\} = (-N\hbar\omega_o/2)\mathrm{Tr}\{\sigma_z\varrho\}$$

where ϱ is the density matrix; and

$$M_z \rightarrow P_z = -W\varkappa/\omega_o.$$

The diagonal energy Hamiltonian operator $\mathcal{H}_o = -\hbar\omega_o\sigma_z/2$ transforms like $I_z = \sigma_z/2$ for $I = \frac{1}{2}$. The quantity P_z, analogous to M_z, appears as a pseudo-electric polarization component. Figure 1 shows the similarity between magnetic and electric dipole two level superposition. Here the pure electronic eigenstates ψ_1 and ψ_2 do not display permanent electric dipole moments. The induced electric polarization

$$P_+ = P_x + iP_y = Np_o\mathrm{Tr}\{\sigma_+\varrho\} = Np_o(u + iv)\exp(-i\omega_o t)$$

is obtained from use of the density matrix ϱ, where $i\hbar\dot{\varrho} = [\mathcal{H},\varrho]$, by applying the transformation

$$\mathcal{H}_r = T^{-1}\mathcal{H}T$$

where

$$T = \exp\{-i(\mathcal{H}_o/\hbar\omega_o)\,[\omega t + \phi(z,t)]\}.$$

This results in the interaction Hamiltonian $\mathcal{H}_r$ matrix (in the "rotating frame")

$$\mathcal{H}_r = -\hbar \begin{pmatrix} \Delta\omega/2 & \varkappa\mathcal{E} \\ \varkappa\mathcal{E} & -\Delta\omega/2 \end{pmatrix},$$

where $\Delta\omega = \Delta\omega_o - \dot{\phi}(z,t)$ and $\Delta\omega_o = \omega_o - \omega$.

A set of polarization components Np_ou, Np_ov, Np_ow appear as slowly varying functions of t and z in a frame of reference rotating at frequency $\omega + \dot{\phi}(z,t)$ where Np_ou is directed along the field modulus $\mathcal{E}$. Therefore

$$\mathbf{P}_r = (\hat{u}_o u + \hat{v}_o v - \hat{z}_o w)Np_o.$$

A schematic of the components of u, v, and W is shown in figure 2. In terms of the components of $\tilde{\varrho} = T^{-1}\varrho T$,

$$u = \tilde{\varrho}_{12} + \tilde{\varrho}_{21},$$

$$v = -i(\tilde{\varrho}_{12} - \tilde{\varrho}_{21}),$$

$$w = -(\tilde{\varrho}_{11} - \tilde{\varrho}_{22}).$$

For small thermal energy $k_BT << \hbar\omega_o$, complete optical ground state occupation gives $w_o \mp \tilde{\varrho}_{11}(0) = -1$, and total ground state energy $W_o = -N\hbar\omega_o/2$. The radiative polarization P_r of the undamped system is coupled to the energy density

$$W = Nw\hbar\omega_o/2$$

according to the conservation condition

$$(Np_o)^2(u^2 + v^2) + [(\varkappa/\omega_o)W]^2 = [(\varkappa/\omega_o)W_o]^2.$$

With damping terms included, the conservation condition no longer applies, and the coupling in general appears in the optical Bloch equations (written for $+|\gamma|$):

$$\dot{u} = \Delta\omega v - u/T_2,$$

$$\dot{v} = -\Delta\omega u - \varkappa\mathcal{E} w - v/T_2,$$

$$\dot{w} = \omega_o\mathcal{E} v - (w - w_o)/T_1.$$

These scalar equations may be given in compact form as

$$\frac{d\mathbf{P}_r}{dt} = \mathbf{P}_r \times [\hat{u}_o\varkappa\mathcal{E}(z,t) + \hat{z}_o(\Delta\omega_o + \dot{\phi})]$$

$$- \frac{(\hat{u}_o u + \hat{v}_o v)Np_o}{T_2} - \frac{(W - W_o)}{T_1}\hat{z}_o.$$

When these equations are coupled to the second order Maxwell's equation in the case of a traveling plane wave propagating in the $+z$ direction (ignoring negligible diffraction effects from sample boundaries), the following first order differential equations result:

$$\frac{\partial\mathcal{E}}{\partial z}(z,t) = -\frac{2\pi\omega Np_o}{\eta c}\int_{-\infty}^{\infty} vg(\Delta\omega_o)\, d\Delta\omega_o, \tag{2a}$$

$$\mathcal{E}\frac{\partial\phi}{\partial z}(z,t) = \frac{2\pi\omega Np_o}{\eta c}\int_{-\infty}^{\infty} ug(\Delta\omega_o)\, d\Delta\omega_o. \tag{2b}$$

where η is the host medium background refractive index. The individual polarizations $Np_o u$ and $Np_o v$ are summed over a normalized inhomogeneous line distribution function $g(\Delta\omega_o)$ where

$$\int_{-\infty}^{\infty} g(\Delta\omega_o)\, d\Delta\omega_o = 1.$$

These equations apply in the slowly varying envelope approximation

$$\frac{\partial\mathcal{E}}{\partial z} << \frac{\mathcal{E}}{\lambda}; \quad \frac{\partial\mathcal{E}}{\partial t} << \varkappa\mathcal{E},$$

where t above corresponds to the retarded time, namely $t_{lab} \rightarrow t + \eta z/c$. Generalization of these equations can be made to cases where the states ψ_1 and ψ_2 are possibly degenerate, and not necessarily characterized by a single dipole matrix element that connects them. If the slowly varying envelope approximation is not made, then terms of the type $\omega\dot{v}$, $\omega\dot{u}$, $\ddot{u}$, and $\ddot{v}$ appear in Eqs. (2a) and (2b) along with dominating terms $\omega^2 v$ and $\omega^2 u$. The size of these higher order terms is usually of the same order of magnitude as the off-resonance response of other quantum levels in the same atom, and these terms are therefore neglected. If the contribution of these higher order terms is to be included, then it is necessary to include the response of other off-resonance quantum levels as well. One may also include the effects of transverse mode character of a non-plane wave, not treated here.

A unique feature of Eq. (2a) shows that any time dependence of ϕ will cause frequency shifts and modulation in any situation where u is not proportional to $\mathcal{E}(z,t)$, and is therefore nonlinear. The effective propagation vector is given by

$$k_e = k + \frac{\partial \phi}{\partial z}(z,t);$$

and the modified frequency is $\omega_e = \omega + \dot{\phi}(z,t)$. Nonlinear deviations from the classical linear field attenuation given by Beer's law

$$\mathcal{E}^2(z) = \mathcal{E}^2(0)e^{-\alpha z}$$

also result from the nonlinear behavior of $Np_o v$ in Eq. (2b), where $\alpha = 8\pi p_o^2 N g(0)\omega/\eta\hbar c$ is the linear absorption coefficient at exact resonance.

IV. ANALOGOUS NMR EFFECTS IN QUANTUM OPTICS

A full review of all the quantum optics experiments that are similar to those known in NMR would take too much space. Instead, mention of the physical importance of a few highlights is the best that this essay can offer. One way of picking out the optical analogs is simply to thumb through the "Old Testament" of *The Principles of Nuclear Magnetism* by Abragam.[18] And even from the "New Testament" on electron spin resonance written by Abragam and Bleaney,[19] for example, the dynamics of phonons interacting with paramagnetic centers in solids can be directly applied to optical photons. (Once at a seminar in Saclay, after I mentioned that Abragam was the only person I knew who was an author of both of the "Old" and "New Testaments," he pointed out that I had overlooked a Higher Being.)

A list is given below of various effects and phenomenology that occur in quantum optics. A good number of the listed items are coincident with those in NMR. Those which do not relate at all to NMR concern fluorescence (spontaneous emission), certain propagation effects, and multilevel Raman processes in unequally spaced multilevel systems. The list does not pretend to be complete, nor is there any assignment of relative importance to each item. Wherever an asterisk (*) is noted, the Bloch formulation has little or no applicability. Although the Bloch-Maxwell equations may be applied directly in describing numerous effects in quantum optics of a semi-classical nature, in some instances where a more rigorous application of the density matrix is required, a resemblance to the Bloch formulation nevertheless may be extracted from the general analysis under certain approximations.

When an asterisk appears with a dagger (*†), it means for certain exceptions the item can be connected usefully with the Bloch formulations, while in most cases it cannot. These assignments are perhaps arbitrary in some cases.

Effects and phenomonology in quantum optics

Linear absorption, dispersion
Nonlinear saturation, hole burning, spectral and spatial diffusion
Adiabatic following and inversion
Rabi oscillation "nutations"
Bloch-Siegert shift*
AC Stark shift*†
Autler-Townes effect*†
Free precession photon (spin-like echo); coherent superposition of quantum states
Double resonance:
  (a) Double excitation with two fields in the laboratory frame of a two level system
  (b) Rotary saturation double excitation with one field in the laboratory frame and one field in the rotating frame of a two level system
Self-induced transparency (manifestation of an electromagnetic soliton)
Coherent radiation "reaction field" damping
Two photon transition dynamics
Stimulated Raman effect*†
Quantum fluorescence beats*
Coherent beats, Raman beats, Raman echoes*†
Raman and Brillouin scattering*
Doppler free Lamb dip and two photon spectroscopy*†
Two level laser Lamb theory
Free electron laser theory
Laser threshold "phase transition"
Fourier transform spectroscopy

Holography*† (in ω-t space the spin-photon echo is an analog of information storage assessed by Fourier analysis in k-r space of optical holography)
Optical self-focusing*
Nonlinear optical pulse chirping, frequency modulation, propagation dependence of phase*†
Nonlinear-optical mixing, harmonic generation, optical rectification*
Homogeneous, inhomogeneous broadening
Relaxation dynamics
Rotating frame polarization locking (production of electric polarization along $\mathcal{E}$ in the rotating frame)

For some of the items listed above it is difficult to make a clear distinction between NMR- and non-NMR-like behavior. For example, in an unequally spaced three level system the Autler-Townes effect concerns the observation of a splitting of a single resonance transition line (connecting level 2 and level 3) into two lines because of the Rabi flop modulation imposed on level 2 by coherent resonance saturation of the $1 \leftrightarrow 2$ transition. This mechanism explains the optical rotary saturation effect[20,21] in a two level system (transition $1 \leftrightarrow 2$) that can be analyzed by the Bloch equations. Figure 3 (p. 33) gives a vector diagram schematic of the rotary saturation mechanism.

The enormous amount of physics implied in the items above cannot be properly outlined here. I hope the reader will have some familiarity with most of them. Wherever the Bloch equations might apply may be seen from the analytical route by the block diagram of figure 4 (p. 34). Various routes of analysis may be taken in the semi-classical (SC) approach. A perturbation approach is often carried out without use of the Bloch equations, given the dotted arrow route. Those not familiar with the direct validity of Bloch equations sometimes take this route by insertion of perturbation source terms into Maxwell's equations directly. Later one may discover that by taking the dashed arrow route, one can include all orders of perturbation simultaneously. This procedure resulted in the discovery of self-induced transparency.[22] The Bloch-Maxwell equations were first solved by computer, and then shown later to be analytically soluble. And finally a limited number of experiments can be analyzed only by quantum electrodynamics (QED), which takes the solid arrow route. These include quantum fluorescence beats, Lamb shift, spontaneous emission, and Dicke superradiance.

Now turn the tables and ask the question: What NMR effects and concepts have not been observed or do not apply in quantum optics, and why? Here is a list of a few of them:

1. Analog of spin temperature in the rotating frame

2. Conserved dipolar energy reservoir
3. Adiabatic cooling
4. Double resonance cross-relaxation coupling between different atomic (spin) species
5. Overhauser effect
6. Solid effect (similar to no. 4)
7. Motional narrowing

In a limited sense one might be surprised to find that items 1, 4, and 6 could be realized in future experiments. In general the analog of NMR "spin temperature" equilibrium in solids does not carry over into the optical case. A single species optical dipolar energy reservoir probably cannot be realized. In a solid, electric dipole-dipole interaction energy is not readily conserved within an isolated two level ensemble of dipoles, a phenomenon that prevents the onset of a quasi-equilibrium temperature different from the lattice temperature. Naturally, such an equilibrium would be possible if the isolated system energy would remain intact for times long compared to the formally estimated dipole-dipole coupling time, designated as T_2. For optical dipoles this equilibrium is not favored because of rapid spontaneous emission, or strong coupling to broad collisional phonon degrees of freedom, which introduce competitive lifetimes comparable to or shorter than the optical system electric dipole-dipole coupling time. (Because of strong coupling, adiabatic cooling and motional narrowing cannot be achieved.) However, a very dilute system of optical dipoles may be assigned a temperature if it is prepared in such a way that it interacts exclusively, for example, by double resonance coupling[20] with an abundant species of spins, where the spins are in dipolar thermal equilibrium contact among themselves. The diagonal susceptibility of the optical system in the laboratory frame provides the coupling mechanism. In the frame of reference rotating with the optical frequency a two level transition occurs in the rf range. The optical polarization appears to be diagonal with respect to the direction of an effective field. The polarization dynamics can be handled by detailed balance or by a set of transformed Bloch equations with relaxation times appropriate to the new frame of reference.[21] A temperature may therefore be assigned to the dilute optical dipoles even though they do not couple directly among themselves. This would require a doped constituent in a solid with a forbidden transition that provides a long fluorescence lifetime.

V. CONCLUSIONS

From this brief summary of quantum optics, we have but one example of the universality of ingenious theoretical formulations by Felix Bloch that

(text continues p. 37)

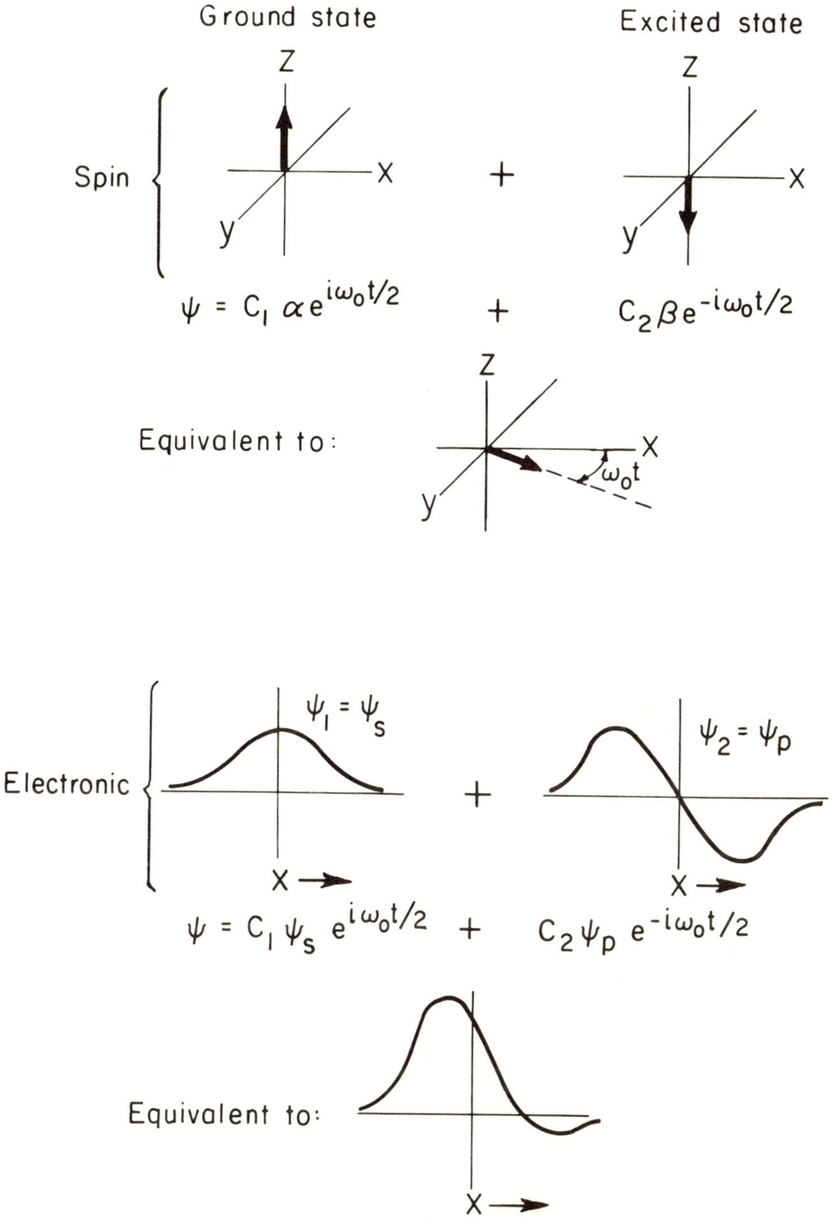

FIG. 1. COMPARISON BETWEEN MAGNETIC SPIN DIPOLE AND OPTICAL ELECTRIC DIPOLE TWO LEVEL SUPERPOSITION STATES. Superposition of spin states $\psi_1 = \alpha$ (spin up) and $\psi_2 = \beta$ (spin down) correspond to superposition of $\psi_1 = \psi_S$ and $\psi_2 = \psi_P$ electronic states, for example. For $C_1 = C_2$, after a 90° pulse, maximum oscillating dipole moments radiate (precess) at the Bohr transition frequency ω_o.

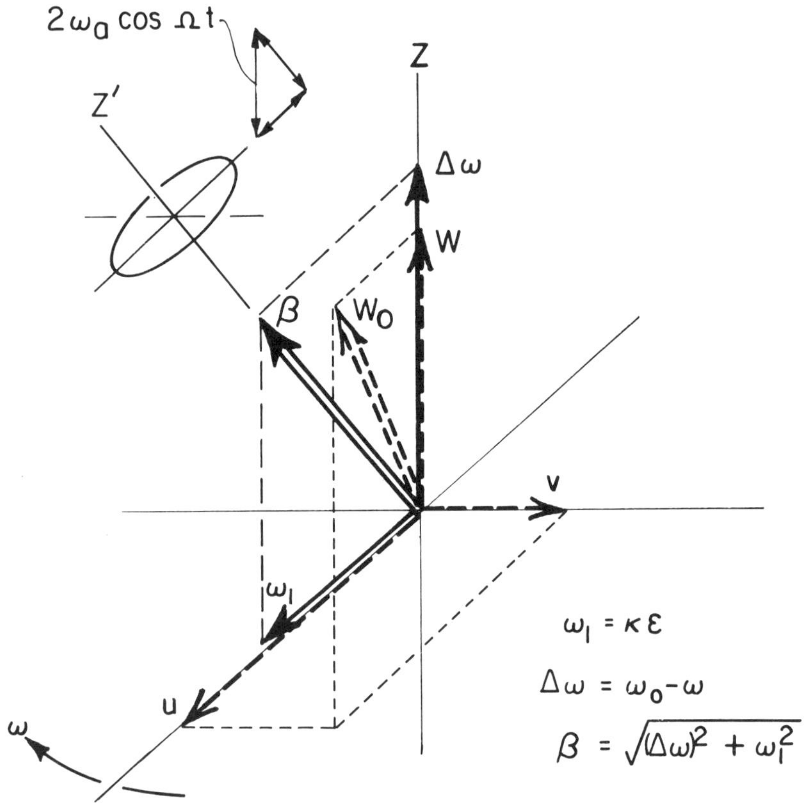

$$\dot{u} = (\Delta\omega + 2\omega_a \cos \Omega t)v - u/T_2$$

$$\dot{v} = -(\Delta\omega + 2\omega_a \cos \Omega t)u - v/T_2 - \kappa\varepsilon W$$

$$\dot{W} = \kappa\varepsilon v - (W - W_0)/T_1$$

FIG. 2. COMPONENTS OF THE VECTOR $\mathbf{P}_r = (\varkappa/\omega_o)\mathbf{W}_o$ in terms of W, u, and v in the reference frame rotating at frequency ω. With damping these components are defined as steady state values from the Bloch equations, where W_o is no longer conserved. If the two level separation is modulated at frequency Ω by an amount $\Delta E = 2\hbar\omega_a \cos \Omega t$ (see figure 3), rotary resonance transitions are induced at frequency $\Omega = \Omega_R = \beta$, and rotary saturation of $\mathbf{P}_r$ occurs. The Bloch equations are shown with the rotary perturbation term included. Here, u, v, and W correspond to u, v, and w (as density matrix elements defined in the text) multiplied by Np_o.

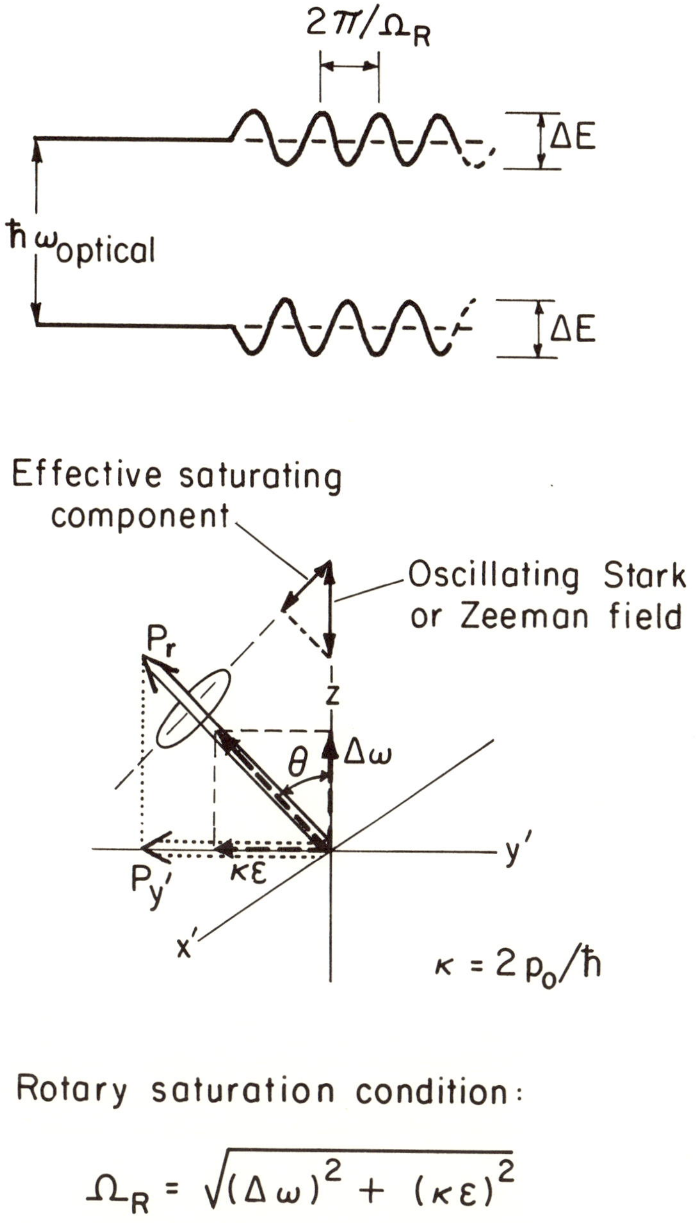

FIG. 3. OPTICAL ROTARY SATURATION OF THE EFFECTIVE POLARIZATION P_r, represented as polarized along the "effective frequency" direction $\boldsymbol{\beta}$ (see figure 2) in the rotating frame.

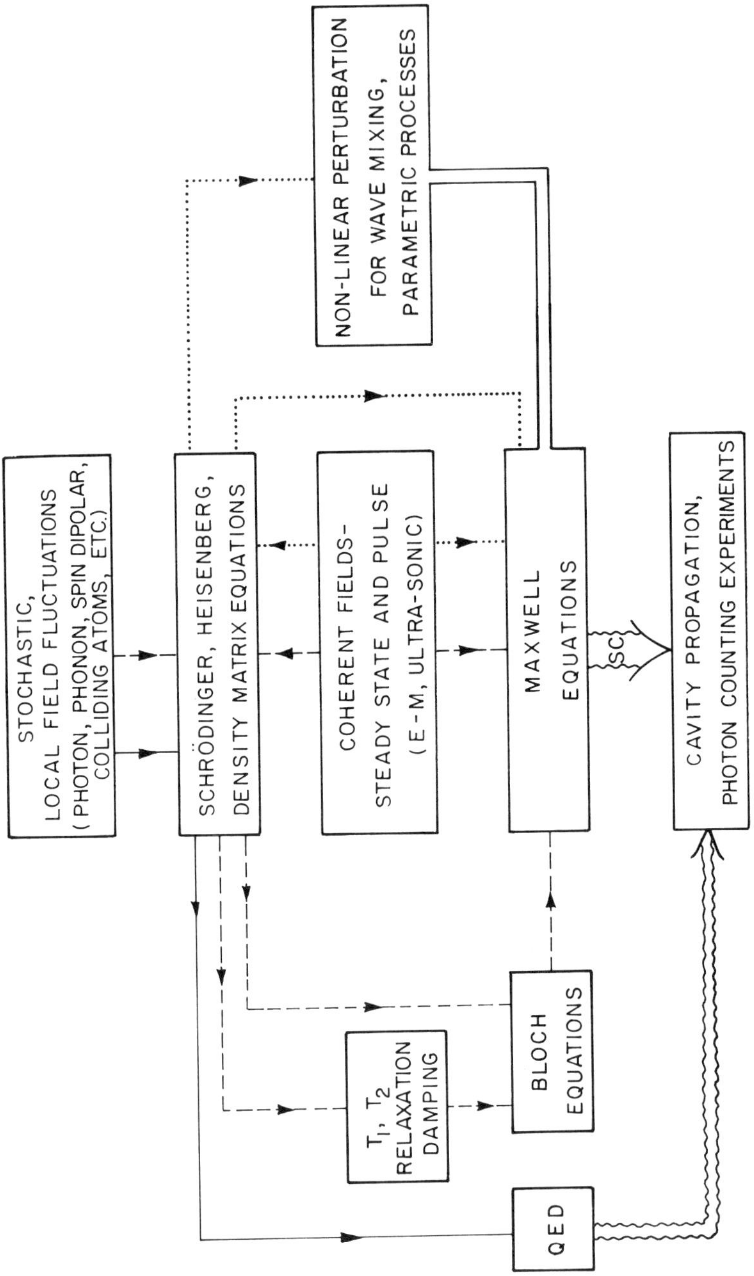

FIG. 4. VARIOUS ROUTES OF ANALYSIS IN QUANTUM OPTICS.

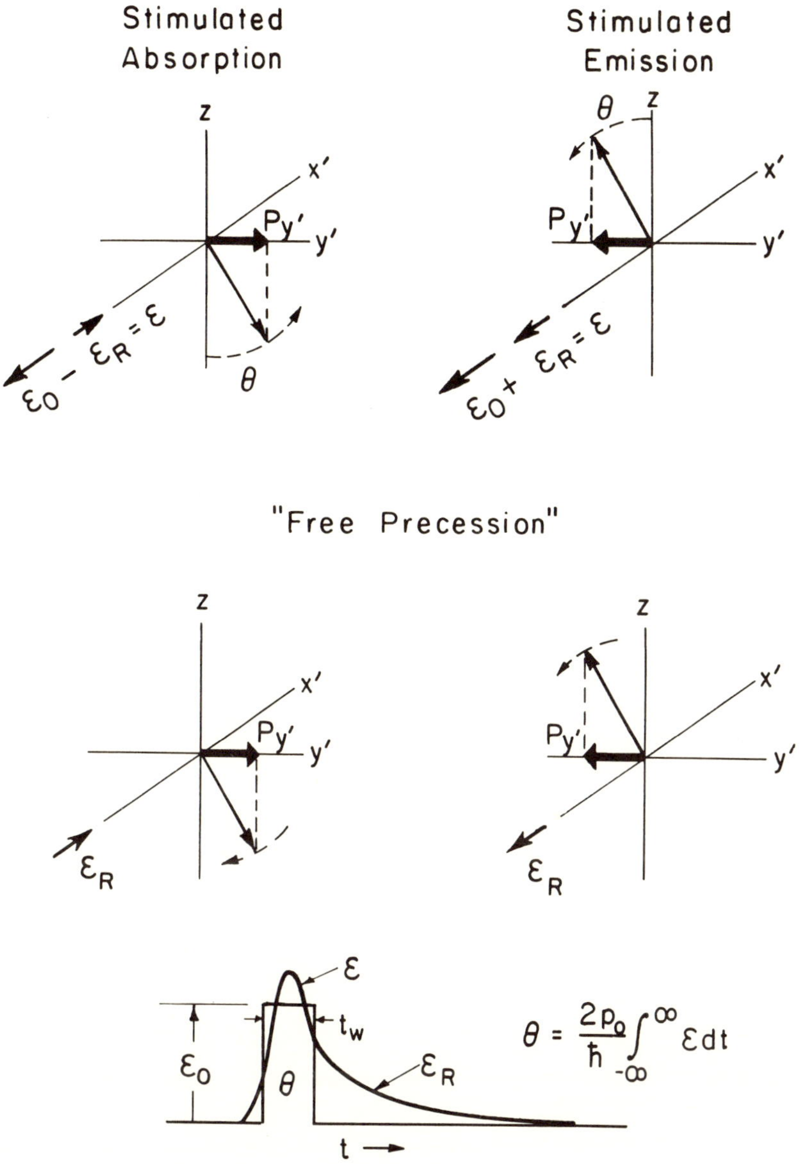

FIG. 5. (top) Stimulated absorption from the ground state and stimulated emission from the inverted state at exact resonance, corresponding to opposite signs of superposition state polarizations $P_{y'}$ in the rotating frame. $\mathcal{E}_o$ is the applied field; $\mathcal{E}_R$ is the reaction field. (bottom) Coherent emission from source polarization $P_{y'}$ after $\mathcal{E}_o$ is suddenly cut off in a time $<< T_2$. Net field $\mathcal{E}$ is a pulse formed from the sum of $\mathcal{E}_o$ (as an applied rectangular pulse) and reaction field $\mathcal{E}_R$. Polarization tipping angle θ is determined by the time integral of $\mathcal{E}$.

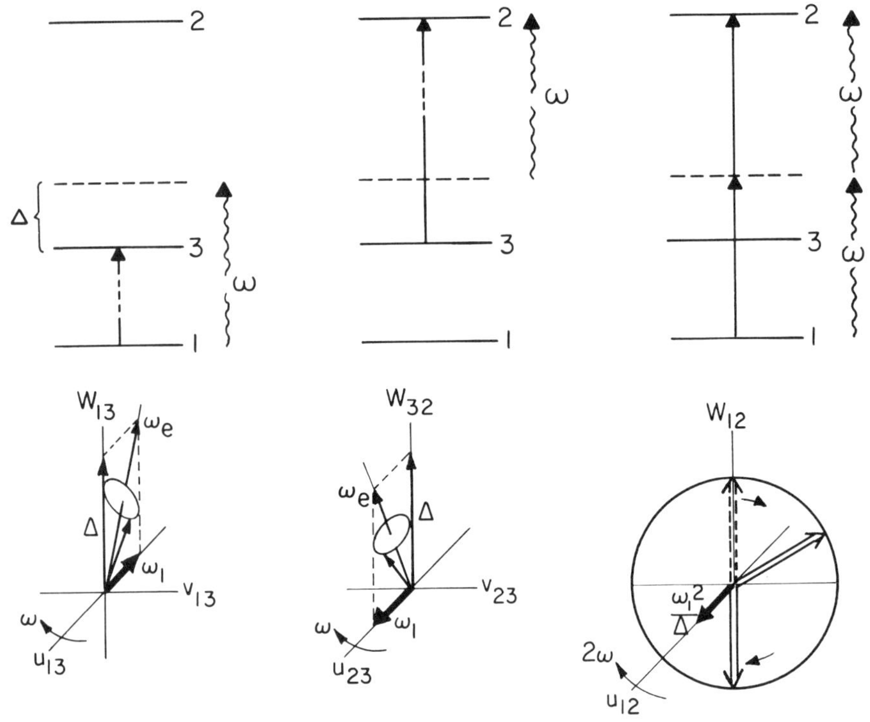

One photon effective field

$$\omega_e = \sqrt{\Delta^2 + \omega_1^2} \sim \Delta \text{ for } \omega_1 << \Delta$$

$$W_{13} + W_{32} = W_{12}$$

Two photon effective field

$$\Omega_e = \omega_1^2/\Delta \approx 1/\hbar^2 \cdot \frac{\langle 1|\mu_{13}H_1|3\rangle\langle 3|\mu_{32}H_1|2\rangle}{\Delta} \text{ for } \delta = 0$$

$$\Omega_e = \sqrt{(\omega_1^2/\Delta)^2 + (2\delta)^2} \text{ for } \delta \neq 0$$

FIG. 6. TWO PHOTON PROCESS compared to one photon processes in a three level system. In terms of NMR, states 1 and 2 are placed in coherent superposition, describable by Bloch equations. The W_{12} polarization precesses at frequency 2ω. The effective frequency Ω_e appears in the rotating frame for two photons interacting slightly off resonance by amount $2\delta << \Delta$. Populations between level pairs are indicated by W_{ij} with corresponding u_{ij}, v_{ij} components. At exact two photon resonance, nutation of W_{12} takes place at the effective Rabi frequency ω_1^2/Δ, where $\omega_1 << \Delta$.

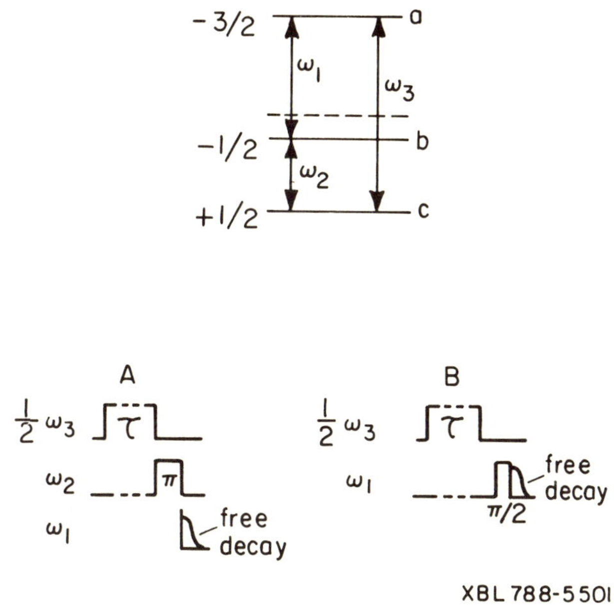

FIG. 7. TRANSFER OF SUPERPOSITION STATE COHERENCE in a three level system with unequally spaced states designated by quantum numbers $m = -3/2$, $-1/2$, $+1/2$ (see ref. 24). A two photon transition with each photon at frequency $\omega_3/2$ is induced by a pulsed field. States a and c are placed in coherent superposition. The far-off resonance conditions $\omega_1 - \omega_3/2 \neq 0$ and $\omega_3/2 - \omega_2 \neq 0$ create negligible direct coupling to state b. No dipole signal is obtained from the $a \leftrightarrow c$ superposition. If a π pulse at frequency ω_2 transfers coherence from level c to b (immediately after the $\omega_3/2$ pulse is shut off), superposition is created between a and b which gives a dipole radiation decay signal at frequency ω_1. Or a $\pi/2$ pulse at ω_1 will give a free precession dipole signal also at ω_1. Conversely, other combinations for coherence transfer are possible. The principle of coherence transfer is useful in coherent Raman phenomena for measurements of relaxation times between levels not directly observable (i.e., $a \leftrightarrow c$).

penetrate many branches of physics. The Bloch equations as they were originally applied to nuclear magnetic resonance crop up in one form or another as a means for solving problems not necessarily related to spin resonance. The Bloch equations have some of the character of the Boltzmann equation, and lend themselves to modification and approximations suitable for the particular problem at hand. As one applies the Liouville equation in a general way, and as the number of elements of the density matrix increases with the number of quantum states, the bewilderment in finding a reasonable solution[23] can sometimes be reduced by seeking Bloch-type equations from suitable combinations of these elements.

It was a personal pleasure and honor to have been affiliated with Felix Bloch in those early days of excitement and enthusiasm when NMR was in its infancy.

ACKNOWLEDGMENT

I wish to acknowledge the support of the Division of Materials of the National Science Foundation.

REFERENCES

1. F. Bloch, Phys. Rev. **70**, 460 (1946).
2. R. Blinc and B. Zeks, *Soft Modes in Ferroelectrics and Anti-ferroelectrics* (North Holland, Amsterdam, 1974).
3. F. A. Hopf, P. Meystre, M. O. Scully, and W. H. Louisell, Phys. Rev. Lett. **37**, 1342 (1976).
4. E. M. Belenov and I. A. Poluektov, Zh. Eksp. Teor. Fiz. **56**, 1407 (1969) [Sov. Phys.-JETP **29**, 754 (1969)]; D. Grischkowsky, M. M. T. Loy, and P. F. Liao, Phys. Rev. **A12**, 2514 (1975); R. G. Brewer and E. L. Hahn, Phys. Rev. **11**, 1641 (1975).
5. E. L. Hahn in *NMR, Basic Principles and Progress*, vol. 13, *Introductory Essays*, ed. M. M. Pintar (Springer-Verlag, Berlin, 1976), pp. 31-44.
6. I. I. Rabi, J. R. Zacharias, S. Millman, and P. Kusch, Phys. Rev. **53**, 318 (1938).
7. N. F. Ramsey, *Molecular Beams* (Oxford University Press, London, 1955).
8. N. Bloembergen, E. M. Purcell, R. V. Pound, Phys. Rev. **73**, 679 (1948).
9. R. H. Dicke, Phys. Rev. **93**, 99 (1954).
10. N. Bloembergen and R. V. Pound, Phys. Rev. **76**, 1059 (1949).
11. A. Szoke and S. Meiboom, Phys. Rev. **113**, 585 (1959).
12. S. L. McCall and E. L. Hahn, Phys. Rev. **183**, 457 (1969).
13. R. P. Feynman, F. L. Vernon, R. W. Hellwarth, J. Appl. Phys. **28**, 49 (1957).
14. E. T. Jaynes and F. W. Cummings, Proc. IEEE **51**, 89 (1963).
15. W. E. Lamb, Phys. Rev. **134**, A1429 (1964).
16. J. P. Gordon, H. J. Zeiger, and C. H. Townes, Phys. Rev. **95**, 282 (1954).
17. N. G. Basov and A. M. Prokhorov, Zh. Eksp. Teor. Fiz. **27**, 43 (1954).
18. A. Abragam, *The Principles of Nuclear Magnetism* (Oxford University Press, London, 1961).
19. A. Abragam and B. Bleaney, *Electron Paramagnetic Resonance of Transition Ions* (Oxford University Press, London, 1970).
20. Yehiam Prior and E. L. Hahn, Phys. Rev. Let. **39**, 1329 (1977).
21. A. G. Redfield, Phys. Rev. **98**, 1787 (1955).
22. S. L. McCall and E. L. Hahn, Phys. Rev. **183**, 457 (1969).
23. R. G. Brewer and E. L. Hahn, Phys. Rev. **11**, 1641 (1975); D. G. Gold and E. L. Hahn, Phys. Rev. **16**, 324 (1977).
24. H. Hatanaka, T. Terao, and T. Hashi, J. Phys. Soc. Jpn. **39**, 835 (1975); H. Hatanaka and T. Hashi, J. Phys. Soc. Jpn. **39**, 1139 (1975).

FIFTY YEARS OF METALS THEORY

by Walter A. Harrison

ABSTRACT

Metallic properties have been attributed to free electrons almost since the discovery of the electron in 1897. This became a theory of metals when Bloch and Sommerfeld applied quantum theory to these electrons in 1928, providing the basis for the subsequent development of the theory. That development was, however, not by a direct process of deduction, but it required genuinely new concepts in order to become simple and understandable. Four such concepts are discussed here: the quasi-particle state of the electron, the representation of its strong interaction with the atoms by a weak pseudopotential, the formation of a local magnetic moment by itinerant electrons, and the superconducting order through which the electrons produce the superconducting state.

I. INTRODUCTION

Just over fifty years ago Felix Bloch published his thesis work on the theory of metals.[1] Viewing from that vantage point, one might feel that the essential problem had been solved and it remained only to work out the detailed consequences. The truth was more that a foundation had been laid upon which a major edifice was to be created. To be sure, many details have been worked out, but I will not discuss them here; I will focus on a series of genuinely new concepts or points of view that were required as the theory moved forward. I begin by describing the status of the theory fifty years ago and noting the signs that were present suggesting undiscovered treasures. We will then see just what some of those treasures turned out to be.

Walter Harrison is Professor of Applied Physics at Stanford University.

The understanding of the properties special to metals began near the turn of the century soon after the electron was discovered by J. J. Thomson.[2] Drude[3] suggested that we could explain immediately some of the most conspicuous properties of metals—their shiny luster and their ability to conduct electricity and heat—if we would assume that the electrons moved freely within a metal and were in thermal equilibrium with the metallic atoms. When Lorentz[4] worked out the consequences of this assumption in the next few years, however, he was led to two serious difficulties. First, the laws of gases required that such an electron gas would need a definite and well-determined amount of energy to raise its temperature—and experiments clearly showed that electrons in metals required very much less. Second, the decreasing resistivity of metals with decreasing temperature required that the electrons must be able to travel increasingly large distances through the metal without colliding with the atoms as the temperature was decreased; there was no basis for this within the theory and indeed it seemed quite implausible. These contradictions remained for some twenty-five years until the arrival of quantum theory.

In a sense the theoretical situation worsened in the interim because of a new experimental development. Onnes[5] explored the behavior of the conductivity of metals at extremely low temperatures to see if indeed the path lengths for the electrons continued to lengthen, with a continuing decrease in resistivity, or whether the electron gas might condense out like frost from the air, giving an increasing resistivity. These two possibilities are illustrated in figure 1. He found that neither was correct. At a critical temperature, T_c, the conductivity surprisingly dropped to exactly zero. We will return later to this remarkable "superconductivity," which remained a mystery for nearly fifty years and was a continual reminder that the theory of metals was not finished.

Quantum mechanics developed in the mid-twenties and immediately the two difficulties raised by Lorentz were resolved. Sommerfeld[6] used the quantum-mechanical statistics for electrons to show that they should *not* absorb energy like a classical gas, and Bloch[1] showed that in a perfect crystal the wave-like electrons of quantum theory could propagate over as long a distance as one wished through the metal. The electron state $|\psi_{\mathbf{k}}\rangle$ could be written as what is now called a *Bloch function*,

$$\psi_{\mathbf{k}}(\mathbf{r}) = u_{\mathbf{k}}(\mathbf{r})\, e^{i\mathbf{k}\cdot\mathbf{r}} \qquad (1)$$

where the factor $e^{i\mathbf{k}\cdot\mathbf{r}}$ describes the propagation of a free electron of wave number $\mathbf{k}$ and the factor $u_{\mathbf{k}}(\mathbf{r})$, which it modulates, has the full translational symmetry of the crystal. It was only with disruptions of the perfection of the crystal, which increase with increasing temperature, that the factors become mixed, the electrons are deflected, and the metal becomes resistive.

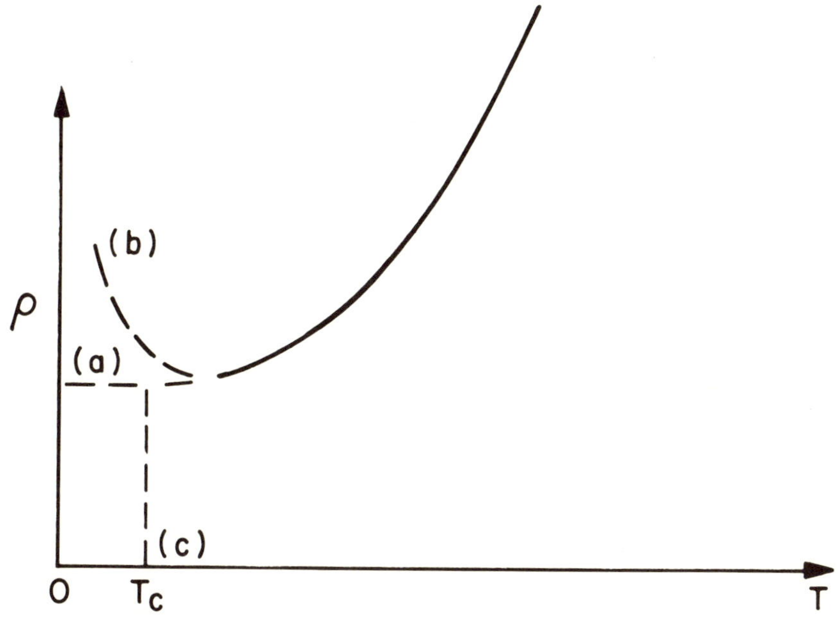

FIG. 1. ONNES IN 1911 MEASURED THE RESISTIVITY ϱ of metals to very low temperature to see if a) it would continue to decrease, or b) it would reach a minimum and increase as the electrons froze out on the atoms. He found instead that, at least for mercury, c) the resistance suddenly dropped to zero, forming a superconducting state.

This finding not only resolved the earlier difficulties, but laid out a path to be followed for the detailed understanding of metals and, in fact, non-metallic solids: One must determine the energies of these wave-like Bloch states of the electrons, and most of the properties of the metals should be understandable in terms of them. Indeed, within the next few years a very extensive theory of metals developed, as reflected in texts by Mott and Jones[7] and by Wilson[8] in 1936.

The theory remained essentially unchanged for the next twenty years, largely because of World War II. In the fifties a number of workers took up the path implicitly laid down in Bloch's work. The newly invented electronic computers allowed the direct theoretical determination of the Bloch states and their energies, studies which are called "energy band calculations." There were still uncertainties as to how these should be carried out, but at the same time pure materials became available and ingenious experimental techniques were developed so that experimental tests could be made of any proposed approximations. This confrontation of experiment and the emerging theory centered on geometrical representations of the directly measurable aspects of the energy bands. These so-called "Fermi surfaces" are characteristic of each metal.[9] By now the Fermi surfaces, and indeed the complete energy bands, are known in some detail for almost all pure metals. The same theoretical techniques should provide the same information for alloys or imperfect crystals and are limited only by the complexity of the systems one is willing to consider.

Such a broad-brush picture of the theory of metals would suggest that there were two great bursts: One followed the discovery eighty years ago of the electron as we learned the consequences of assuming that the electron in the metal was subject to the classical laws of mechanics and statistical mechanics. The second followed the development fifty years ago of the quantum mechanics and quantum statistics that, in fact, describe the behavior of electrons. If we look more closely at the past fifty years we see that a number of additional genuinely new concepts were required, each followed by its own careful exploration. It is these new concepts in the past fifty years that I wish to discuss, rather than the very impressive consequences that have derived from them. This then is a study of the development of a field of science in which the fundamental laws appeared to be known.

I have implicitly made a distinction between two kinds of new concepts in science: creations, such as the invention of quantum theory, and discoveries of existing entities, such as electrons. We see in the theory of metals that both kinds of steps are necessary and there seems to be a sharp distinction between them. We should consider the difference more closely.

Perhaps a very simple example of invention in a known field would be helpful. We might seek the description of the motion of a string of beads hurled through the air or through a vacuum, as illustrated in figure 2. If we

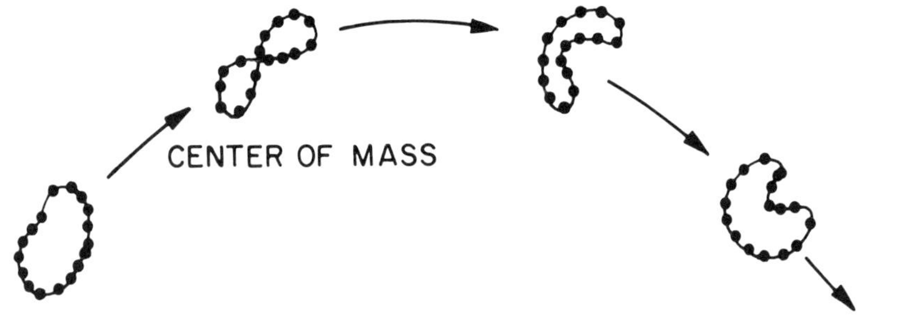

FIG. 2. THE DYNAMICAL LAWS GOVERNING THE TRAJECTORY of a string of beads are known, as are the laws that govern electrons in metals. Simultaneous solution of the equations of motion for each bead is a correct procedure, but the concept of the "center of mass" allows one easily, and reasonably accurately, to predict where the necklace will land.

specify the velocity of each bead at the beginning of the trajectory, we may calculate the forces of the beads on each other and the gravitational force on each and could calculate the subsequent motion; the laws of the dynamics are known. However, we might invent or define the *center of mass* of the necklace. We may immediately then show that the motion of the center of mass is very simple, and by considering just its motion we may be able to calculate on the back of an envelope where the necklace will land. We have given up some accuracy in doing this, since we do not tell just where each bead lands, but if the necklace has been thrown a long way, it may be enough to know where the necklace as a whole hits the ground. Without that invention, the solution of the equations of motion for each of the beads would have required a sizable computer. The concept of "center of mass" was never necessary to the solution of the problem, but its definition was necessary to extract the simplicity from an intrinsically complex situation.

One could argue, on the other side, that a concept such as the center of mass (or the other concepts I will discuss) so naturally and simply describes the physical world that if one person did not invent it someone else would. In this sense the invention is essentially a discovery of a concept waiting to be found. It may be a semantic point of distinction, but I hope the illustration does clarify the parts of the theory I wish to discuss.

Before beginning, I should note where the necessity of the new concepts could be spotted in the early theory. I noted that Bloch's theorem about the motion of the electrons depended upon the perfect crystalline symmetry of the metallic lattice, but Mott[7] noted very early that the resistivity of most metals does not change greatly upon melting, at which time this periodicity is lost altogether. Bloch's theorem is unquestionably true, but is there some more general reason that applies to much more general systems? Furthermore, the Bloch states were derived by consideration of a single electron moving through the metal. Any estimate would suggest that the electrons in metals are in continual contact with each other (through their electrostatic interaction) like marbles in a bag or molecules in a liquid; how can we then describe them as moving in independent trajectories through the metal? Indeed, when Bloch and others went on to describe detailed properties of ferromagnetic metals, they thought of the electrons as little magnets attached to individual atoms, as if they had frozen out, as Onnes speculated. This was quite inconsistent with the original Bloch states. How can the two views be reconciled? Finally, there was the problem of superconductivity, which remained outside the domain of respectable theory of metals and was scarcely mentioned in the earlier texts.[7,8] We will consider concepts that in some sense made these very difficult problems simple.

II. THE ELECTRON-ELECTRON INTERACTION

It is good to begin with the problem of the mutually interacting gas of electrons in a metal, which is truly an insoluble problem, though the needed fundamental laws are precisely known. It is well known in classical physics that one can solve generally for the motion of two interacting particles, but not for three. For three particles one must resort to numerical or approximate methods. The same is true in quantum mechanics, and with a typical metal crystal containing 10^{22} electrons, the direct numerical methods are out of the question and approximational techniques need to be developed. This branch of physics is called "many-body physics" and is concerned with other problems as well. I would like to describe an approach to the problem that transcends the detailed solution, much as the center of mass did for the beads. L. D. Landau,[10] in the Soviet Union, suggested that we should first imagine a system of particles (electrons) that did not interact with each other at all. Then each electron would have a well-defined wave number. We represent the states by a wave number as in Eq. (1), though Landau was not concerned about the effect of the crystal potential. He therefore regarded $u_{\mathbf{k}}(\mathbf{r})$ in Eq. (1) as a constant. In this case $\mathbf{k}$ becomes the momentum divided by Planck's constant, $\mathbf{k} = \mathbf{p}/\hbar$, which would not change because the electron did not interact with any other electrons. The state of the entire system could be given by listing (for each of the two possible electron spin orientations) all possible wave numbers, indicating whether or not an electron occupied each corresponding state; i.e., we give the function $n(\mathbf{k})$, with $n = 1$ if the state is occupied and $n = 0$ if it is not. (We let $n(\mathbf{k})$ represent the occupation of both spin states.) Landau then asserted that even if the electrons interact with each other, the resulting state of the interacting system can again be completely specified by an $n(\mathbf{k})$, which he called the occupation of *quasiparticle* states. It follows that the total energy of the system of real electrons is a *functional* of the function $n(\mathbf{k})$; that is, $E\{n(\mathbf{k})\}$. Thus he asserted that the entire effect of the interactions was to replace the freely propagating electron by a quasiparticle consisting of the electron and a "wake," which it carries with it as it moves through the sea of other electrons. Then the quantity in the interacting system that plays the role of the electron energy in the noninteracting system is the quasiparticle energy $\epsilon_{\mathbf{k}}$ defined as the change of the total energy when an electron is added in the quasiparticle state $|\mathbf{k}\rangle$:

$$\epsilon_{\mathbf{k}} = \delta E\{n(\mathbf{k})\}/\delta n(\mathbf{k}) \qquad (2)$$

The interactions affect the energy and the velocity of the quasiparticle, but these changes might be obtained from experiment, and the concept allowed us to continue to think simply about an intrinsically complex, and in fact insoluble, problem.

When this theory reached the West (there was a delay of some years because translations were not made routinely during the Cold War), it was regarded as a rather bold, and perhaps unjustified, assumption. Efforts were initiated to see if the Fermi liquid theory could be derived from the rigorous and powerful "quantum field theories," which had been developed here and which were the basis of many-body physics in the West. Landau's response was interesting; he thought it would be a nice test of the expansions used in quantum field theory. This was not the outrageous comment it might at first seem. Quantum field theory began with the noninteracting electrons and systematically corrected for the interaction. It necessarily, therefore, led to a state of the system that could be specified completely by giving the $n(\mathbf{k})$ of the zero-order (noninteracting) state and therefore satisfied Landau's starting assumption. Landau might conceivably have been wrong with his Fermi liquid theory, but if so the perturbation theory used in quantum field theory was necessarily also inapplicable. We will see that both are inapplicable in superconducting metals.

One thing Fermi liquid theory could *not* do, which field theory could, was predict the size of the corrections to energy and velocity for the quasiparticles. It was difficult, but efforts by theorists to calculate the corrections, along with concurrent experimental studies of metals, made clear that the dynamical properties of the quasiparticles were only slightly different from those of noninteracting electrons, and the independent-electron view used by Bloch was justified. This did not solve all of the problems. Even if electrons could be studied one at a time, we must know how to approximate the potential they move in if we are to determine the electron (or quasiparticle) states and energies. Procedures for constructing these have come through hard work and experience.

III. ELECTRON-ATOM INTERACTIONS

I turn next to the interactions between the individual electronic quasiparticles and the atoms that make up the crystal, and why they can travel so far in the metal even if the crystalline symmetry necessary to Bloch's theorem is absent. Before it was understood why there was such a small change in resistivity when a metal melted, it was learned that the energies of the Bloch states in perfect crystals were very little different from what they would be if the electrons could move completely freely through the metal, as Drude had originally assumed. This had been learned from experimental studies of metal Fermi surfaces,[9] and it also followed from the band calculations of the states. From the theory one could see that the effect of the potential arising from the atoms was small because of the presence of deep electronic states, core states $|c\rangle$, on each atom, which were otherwise of little interest. In fact a rather good approximation to the Bloch function $|\psi_{\mathbf{k}}>$ can be obtained from the free-electron plane-wave state $|\mathbf{k}>$ simply by orthogonalizing it to the core states,

$$|\psi_{\mathbf{k}}> \; = \; |\mathbf{k}> - \Sigma_c |c> <c|\mathbf{k}> \tag{3}$$

Corrections to this "orthogonalized plane-wave state" then could be calculated in terms of a *pseudopotential*, W(**r**), which included not only the effect of the potential $V(\mathbf{r})$, arising from the atoms, but also a cancelling effect from the orthogonality,

$$W(\mathbf{r}) \; = \; V(\mathbf{r}) + \Sigma_c(\epsilon_{\mathbf{k}} - \epsilon_c)|c> <c| \tag{4}$$

The pseudopotential entered the theory of the Bloch electrons just like the true potential, but it was quite weak, as is illustrated in figure 3.

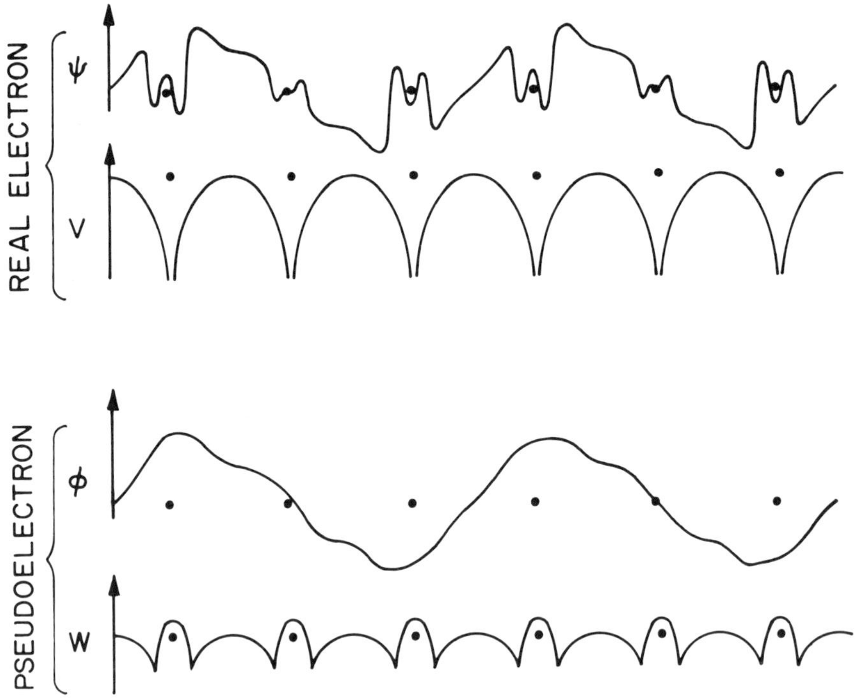

FIG. 3. IN THE UPPER FIGURE the potential V seen by the electrons in a metal is sketched, along with the Bloch state ψ, which describes the electron in the metal. In the figure below is a pseudopotential W, which differs from the potential V only near the nuclei. It is constructed so that the corresponding pseudowavefunction ϕ is exactly equal to the Bloch state except near the nucleus. The pseudowavefunction is simpler than the Bloch state, since it derives from a weak pseudopotential, but its energy and dynamical properties are the same.

The idea of a pseudopotential was not new. Fermi had first introduced it in the theory of collisions between protons and neutrons; however, that problem was sufficiently simple that the concept was not really needed. It had also been used by Phillips and Kleinman[11] to simplify band calculations for semiconductors. It is interesting that for semiconductors the smallness of the pseudopotential was not of importance, but the concept was useful in suggesting approximations that made the band calculation simpler without significant loss of accuracy. It could be used in a similar way to simplify the band calculations of metals, but in metals the smallness was of real utility.

Because the pseudopotential is weak, one may begin with the free-electron theory of Drude, using the quantum statistics as Sommerfeld had done, and directly calculate the corrections due to the pseudopotential. This made it immediately possible to estimate the entire range of properties of the metal,[12] including the electrical resistivity of the liquid metal. That resistivity came out to be very small, in accord with experiment, reflecting the weakness of the pseudopotential that describes the interaction between the electrons and the atoms in the metal, whether it be crystalline or liquid.

Pseudopotential perturbation theory was not limited to electrical properties, but allowed the calculation of the total energy of the metal as a perfect crystal, a distorted crystal, or one with crystalline defects. Thus it allowed calculation of the elastic properties, the vibration spectrum, and the energy of formation and properties of defects. This was the first state of condensed matter for which there existed a theory allowing the calculation of essentially all properties in terms of the basic electronic structure.

The essential step in this advance was the recognition that the effect of the electron-atom interaction could be used mathematically as a small correction, the concept of a weak pseudopotential. Theories have subsequently been developed that allow the calculation of the properties of most other types of solids in terms of the electronic structure.[13] It is interesting, however, that only in the case of the simple metals is the mathematical treatment of the ratio of the pseudopotential to the kinetic energy as the small parameter appropriate; in semiconductors, where pseudopotentials had been introduced earlier,[11] just the reverse is required, an expansion in the ratio of kinetic energy to the pseudopotential. The pseudopotential is a completely valid concept in all systems, but it is in the simple metals that it reduces an extremely intricate problem to one that is quite simple to understand and to calculate.

IV. MAGNETISM

Magnetism is a familiar property of special metals, such as iron, and at the time of Bloch's thesis it was recognized that it arose from the spin of the metallic electrons. Bloch in fact showed[14] that in a standard approximation (the Hartree-Fock approximation) an electron gas at low enough density would become ferromagnetic. The physics of this analysis was essentially correct. Though the electronic structure in ferromagnetic metals is always much more complicated, his calculation can be redone using Bloch states for the metal crystal; it gives a good account of the perfect ferromagnetic state. This was first done in the sixties for nickel by Wakoh and Yamashita[15] and by Connolly.[16]

As in the problem of resistivity discussed in section III, the difficulty comes with the disordered system. If iron is heated above a critical temperature, the ferromagnetism is lost but the metal does not return to the usual nonmagnetic state. The magnetic properties are much like what would be expected if the electrons had frozen out on the atoms as Onnes had speculated[5] and the ferromagnetic state consisted of aligned electron spins on each atom that became disordered above the critical temperature. Such a disordered array of spins is quite inconsistent with Bloch states for the electrons. It suggested that in the ferromagnet the electrons must be localized, existing in qualitatively different kinds of states from the itinerant electrons of Drude and Bloch. This provides a simple description of ferromagnetism, but one which turns out not to apply to metals such as iron and nickel; measurements of the Fermi surfaces of these materials show "up-spin" and "down-spin" Fermi surfaces and little doubt that the electrons are itinerant. The paradox was resolved by Anderson's introduction of a *local moment* arising from itinerant electrons.[17] (This should not be confused with another well known part of Anderson's work concerning the true localization of electronic states in disordered systems.)

Anderson considered a single atom, such as iron, dissolved in a simple metal, such as copper. The magnetic electron states, $|d\uparrow\rangle$ or $|d\downarrow\rangle$ depending on spin direction, from the iron atom cannot freeze out an electron from the metal. They become what are called "resonant states," in which a passing electron may linger before moving on, just as a shopper may take one turn around a revolving door on the way into a store. This behavior is illustrated in figure 4. The analogy may be carried further to illustrate the local moment: The shopper could have gone to the right (counterclockwise) through the door, or to the left. A second shopper coming immediately after will find it much easier to go the same way as his predecessor. Thus for an extended period of time the door may be moving continually to the right as the shoppers flow in and out of the store, each taking one turn around on the way through. If we do not notice the difference between different shoppers, we may think one shopper is going round and round in a localized state as the others pass through.

The analogy with the local moment is immediate. An electron passing by may "resonate" at the iron atom and then move on. It might be an

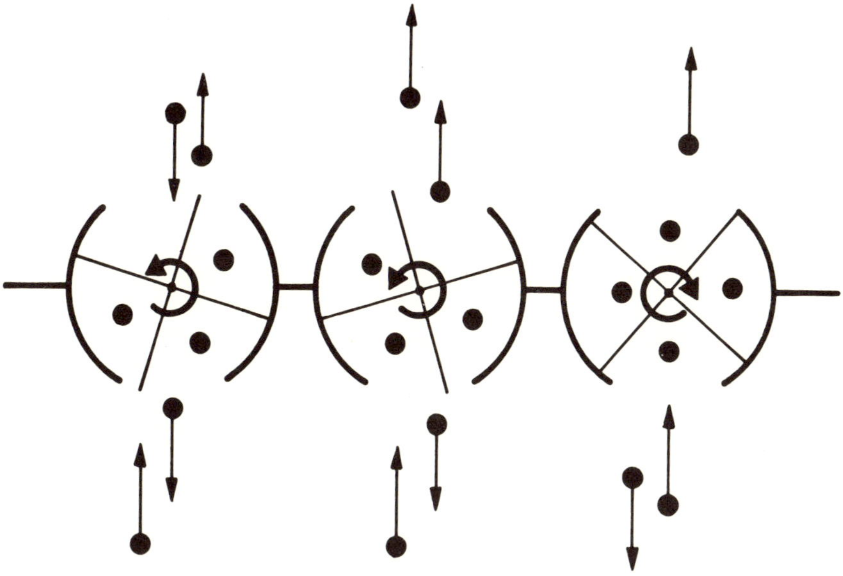

FIG. 4. THE FORMATION OF LOCAL MAGNETIC MOMENTS by itinerant electrons near an iron atom is analogous to the flow of people through a revolving door. Each may take a turn or two around on the way through, and if you cannot tell them apart, there is the same effect as if individuals were trapped there and the others went right through. Furthermore, if the first one goes to the right, corresponding to a magnetic moment directed into the plane of the figure, those that follow will tend to go the same way. In this way a ferromagnetic metal atom can acquire a long-lasting magnetic moment without holding any one electron an appreciable time.

electron of up-spin or down-spin, but if it is of up-spin, the next electron to resonate will more likely be up-spin (by the same physical mechanism that caused all electrons in Bloch's low-density electron gas to have the same spin), so that a local up-spin density is produced by the multitude of passing electrons. Since electrons are fundamentally indistinguishable, this is magnetically the same as if an up-spin electron had been condensed onto the atom, although the electrons' transport past the atom is as if they were all free.

An approximation to the resonant electron state can be made by adding some terms from the *d*-states of the free atom to each free-electron state. The form in which these enter, for electrons of up-spin for example, is

$$|\psi_{\mathbf{k}\uparrow}\rangle = |\mathbf{k}\uparrow\rangle + \frac{|d\uparrow\rangle\langle d\uparrow|V|\mathbf{k}\uparrow\rangle}{\epsilon_{\mathbf{k}} - \epsilon_{d\uparrow}} \qquad (5)$$

The corresponding expression can be written immediately for down-spin. However, if this happens to be an up-spin resonance, the energy $\epsilon_{d\uparrow}$ is lower than $\epsilon_{d\downarrow}$ because of the other electrons of up-spin. The low value of $\epsilon_{d\uparrow}$ in Eq. (5) may enhance the resonance of this electron at the iron atom, cooperatively lowering the $\epsilon_{d\uparrow}$ for the other electrons; it adds its push to the revolving door.

This resonant electron state is not truly a Bloch state unless we construct a periodic array of such states with the same spin. Then each electronic state would have a well defined direction of spin, as in Bloch's states. However, we could make the corresponding array of iron atoms as far apart as we please and it is clear that the behavior of the system is that of rather independent "local moments" at each atom. If we think then in terms of these local moments, it is easy to understand how they may be disordered. We can, in fact, calculate the dependence of the energy upon the relative direction of moment on two atoms, the so-called "Ruderman-Kittel[18] interaction," which ultimately produces the ferromagnetic state once the iron atoms are quite close together.

The concept of the local moment enables us to transcend a difficult theoretical problem. The concept is extracted from the theory of an isolated spin and then used directly to study the multiple-moment system without worrying about the fact that the final disordered state we wish to discuss does not contain electron states of well-defined spin. Perhaps it is analogous to extracting the concept of "center of mass" from the string of beads and then calculating its simple parabolic trajectory although no bead is in itself undergoing such simple motion. We may even worry in that case about some approximation we made in specifying the interaction between different beads, but the center of mass transcends these details.

V. SUPERCONDUCTIVITY

I turn finally to the nature of a superconductor, which defied explanation for nearly fifty years after its discovery. It is interesting in the context of our discussion in section II that Landau and his colleague Ginsburg noted that an additional parameter, *the superconducting order parameter*, was needed to describe the superconducting state.[19] This violated the conditions necessary for the Fermi liquid theory, developed several years later and discussed in section II, to apply. This order parameter had a role in the theory of superconductivity very much like the role of spin ordering in the theory of ferromagnetism, but in the case of superconductivity the Ginsburg-Landau theory did not tell what the physical nature of this superconducting order was. It was known from other experiments (the isotope effect) that the origin of the state was related to the vibrations of the metal lattice, but the failure of the condition for the Fermi liquid theory would suggest, correctly, that the superconducting state could not possibly be found by the perturbative techniques of quantum field theory. Perhaps this is why the nature of the state was so elusive. When it was finally revealed by Bardeen, Cooper, and Schrieffer,[20] the mathematical form of the energy in fact could not be written as an expansion in the interaction, confirming that the result was unobtainable by treating the interaction in perturbation theory.

The form of this BCS theory of superconductivity is intrinsically quite mathematical and difficult, though in the hands of the able theorist it is very accurate and very complete. Here we will discuss only one concept that makes it simple by transcending the details of the theory: this is the superconducting order parameter, introduced by Ginsburg and Landau, the nature of which was learned from the BCS theory. We do this first by analogy with a more familiar problem, that of radio-wave radiation from an antenna.

If we were to describe the radio antenna in terms of quantum theory, we would say that it emitted quanta of radiant energy, called photons, in all directions, as illustrated in figure 5. Then our home receiver is a photon

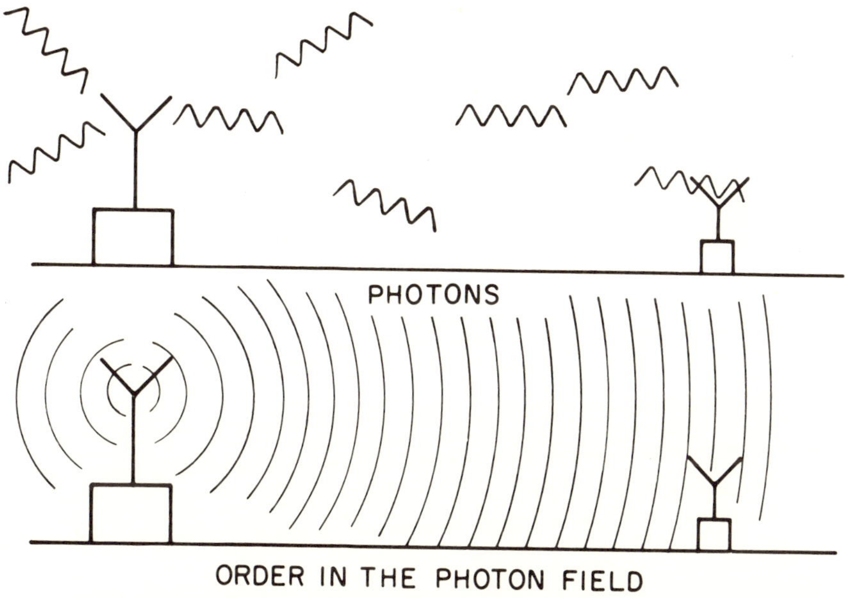

FIG. 5. IN QUANTUM THEORY a radio antenna can be considered as an emitter of photons that are collected by the radio receiver as shown in the upper figure. A slightly less accurate, but much simpler, description is obtained by introducing an electric field depending upon position and time, illustrated below. This electric field may be considered an order parameter in the photon field, and its definition requires one to abandon the idea of strict conservation of the number of photons.

collector, which detects the photons that come its way, unscrambles the time variation of the photon density, and produces the sound. This is a rigorous and exact treatment of radio communication, but, like the equations of motion of each bead, it is an extremely difficult way to get the answer. Any electrical engineer will instead define an electric field that varies in time and space and use Maxwell's Equations to estimate the resulting field on the home receiver, and finally he will deduce the sound intensity coming from the radio. What this engineer has done, in the context of our discussion here, is to introduce an order parameter (the electric field) in the gas of photons coming from the antenna. This approach is slightly less precise than the quantum theory, since in order to define the phase of the electric field we must give up the possibility of specifying the number of photons present, or the exact energy in the radiation field (a direct consequence of the Heisenberg Uncertainty Principle).

A more mathematical way of saying the same thing is to describe the state of radiation by specifying the number of photons of each wave number, $n(\mathbf{k})$, just as we specified the number of electrons (zero or one) in each electron state. Then we imagine a "creation operator" $a^+_\mathbf{k}$, which is defined so that the state $a^+_\mathbf{k}|n(\mathbf{k})>$ has $n+1$ photons in the state $\mathbf{k}$. Since $a^+_\mathbf{k}|n(\mathbf{k})>$ and $|n(\mathbf{k})>$ are different states of the photon system, one says that the average value of the operator $a^+_\mathbf{k}$ is zero. The electric field turns out to be proportional to the creation operator for photons, so it can have a nonzero average only if we discuss a state that has some probability of containing n photons and some probability of $n+1$ photons; that is, a state with an ill-defined number of photons. The state of the system must in fact be written as a sum, with coefficients u_n, of states $(a^+_\mathbf{k})^n\ |0>$ with many different numbers n of photons:

$$|\psi_\mathbf{k}> \ = \ \Sigma_n u_n (a^+_\mathbf{k})^n |0> \tag{6}$$

The u_n may be small except for n near the average value $\bar{n}$, but it must be nonzero over some range of n.

At first we might worry that we must talk of a state of uncertain energy, but on second thought we realize that the energy of the field from an antenna is so large that the uncertainties become negligible in comparison. It is customary under such circumstances to say that this is the classical limit or that quantum effects are negligible. This is the exact analogue to calculating the trajectory of a ball (or the necklace) and noting that if we specify exactly the position of the center of mass, the momentum becomes uncertain, but in the classical limit the uncertainty is negligible.

In the case of the superconductor, we learned from the BCS theory, and work following from it, that the order parameter is proportional to the product of two electron creation operators, one for electrons of up-spin and one for electrons of down-spin. In its lowest energy state these electrons are to be moving in opposite directions so that, in analogy with Eq. (6), the BCS ground state uses a sum of terms $v_k a^+_{-k\downarrow} a^+_{k\uparrow}$ and u_k corresponding to the pair of states being occupied or empty, respectively. The BCS ground state is a product of such factors for each wave number,

$$|\psi_{BCS}\rangle = \Pi_k(u_k + v_k a^+_{-k} a^+_k)\,|0\rangle \tag{7}$$

The implications of this remarkable finding follow from the analogy with the antenna. We note first that the problem becomes much simpler because we relaxed any requirement of a fixed number of electrons. This is somehow much more difficult to accept than the relaxation of the requirement of a fixed number of photons, but it is much the same. For a reasonably sized superconductor, the uncertainty is extremely small compared to the total number of electrons. If two metallic specimens were connected so that electrons could flow between them, the number of electrons in each system would actually have corresponding uncertainties, and we know experimentally that it would not noticeably change the properties.

From this point of view the superconducting order parameter becomes a macroscopic classical variable, analogous to an electric field, which in this case is observable through superconducting currents that can be written in terms of it. In this regard also it introduces another variable in our electrical circuits; that variable is the phase difference of the order parameter that directly determines the current flow through a Josephson junction and must be included along with the voltage in solving for the current in the circuit. It is also a system parameter like the magnetization of iron that specifies the state of the metal.

Finally, it is fair to ask for the microscopic origin of the superconducting order, the analogy to the spin alignment in the case of ferromagnetism. The fact that it derives from a product of two creation operators, for up-spin and down-spin, means that it arises from pairs of electrons of opposing spins, called Cooper pairs. This clue to the origin came before the BCS ground state was found; Cooper[21] showed that the normal state of the metal could be lowered by forming one such pair; the difficulty then came in seeing that to make many Cooper pairs was simple only if electron conservation was relaxed. Then as long as the center of mass of each pair drifted at the same rate, it was possible to pair up essentially all of the electrons, with the electrons of each pair orbiting one another at such a large distance that each orbit crosses many others. Schrieffer[22] makes the analogy to pairs of dancers with each keeping track of his partner though they move in quite different parts of a crowded ballroom. This is possible for Cooper pairs only if the center of gravity of every pair is stationary, or if they all drift together. In either case the coherent motion, or lack of motion, of the Cooper pairs is exactly what is represented by the order parameter. It is also what prevents a superconducting current, once set in motion, from being dissipated by individual electrons colliding with defects in the crystal (see figure 6).

As with the center of mass, the quasiparticle, the pseudopotential, and the local moment, the superconducting order parameter is not necessary to the understanding of the phenomenon, but is an insight that can suddenly make the phenomenon seem simple and therefore understandable. Whether these concepts be inventions or discoveries, they are, in my view, the highest achievements of the theory of metals in the fifty years since Bloch's introduction of Bloch states to the theory.

ACKNOWLEDGMENTS

Edward L. Ginzton Laboratory Report No. 2826, supported in part by NSF Grant DMR 77-21384.

REFERENCES

1. F. Bloch, Z. Phys. **52**, 555 (1928).
2. J. J. Thomson, Philos. Mag. **44**, 293 (1897).
3. P. Drude, Ann. Phys. (Leipzig) **1**, 566 (1900).
4. H. A. Lorentz, *The Theory of Electrons* (Teubner, Leipzig, 1909).
5. H. Kamerlingh Onnes, Commun. Phys. Lab. Univ. Leiden, Nos. 119-122 (1911).
6. A. Sommerfeld, Z. Phys. **47**, 1 (1928).
7. N. F. Mott and H. Jones, *The Theory of the Properties of Metals and Alloys* (Clarendon Press, Oxford, 1936).
8. A. H. Wilson, *The Theory of Metals* (Cambridge University Press, Cambridge, 1936).
9. *The Fermi Surface*, ed. by W. A. Harrison and M. B. Webb (Wiley, New York, 1960).
10. L. D. Landau, J. Exptl. Theoret. Phys. (U.S.S.R.) **30**, 1058 (1956).
11. J. C. Phillips and L. Kleinman, Phys. Rev. **116**, 287 (1959).
12. W. A. Harrison, Phys. Rev. **129**, 2503, 2512 (1963).
13. For a complete discussion see W. A. Harrison, *Electronic Structure and the Properties of Solids* (W. H. Freeman, San Francisco, 1980).
14. F. Bloch, Z. Phys. **57**, 545 (1929).
15. S. Wakoh and J. Yamashita, J. Phys. Soc. Jpn. **21**, 1712 (1966).
16. J. W. D. Connolly, Phys. Rev. **159**, 415 (1967).
17. P. W. Anderson, Phys. Rev. **124**, 41 (1961).
18. M. A. Ruderman and C. Kittel, Phys. Rev. **96**, 99 (1954).
19. V. L. Ginsburg and L. D. Landau, J. Exptl. Theoret. Phys. (U.S.S.R.) **20**, 1064 (1950).
20. J. Bardeen, L. N. Cooper, and J. R. Schrieffer, Phys. Rev. **108**, 1175 (1957).
21. L. N. Cooper, Phys. Rev. **104**, 1189 (1956).
22. J. R. Schrieffer, private communication.

ORDER IN A SUPERCONDUCTOR

FIG. 6. THE SUPERCONDUCTING STATE arises from a subtle ordering of the electrons in which they form pairs rotating around each other, so that the center of mass of every pair drifts exactly together. In the figure each empty circle is at the opposite diameter of its orbit as a filled circle moving in the same orbit. Thus if a current is flowing, the collision of one electron with an obstacle, here an impurity drawn as a square, will cause its partner to be deflected also, without reducing the total flow of current. The definition of the corresponding order parameter requires the relaxation of strict electron conservation in analogy with the introduction of the electric field in figure 5.

NUCLEAR MAGNETIC RESONANCE AS A STRUCTURAL METHOD IN MOLECULAR BIOLOGY

by Oleg Jardetzky and Norma G. Wade-Jardetzky

ABSTRACT

The discovery of nuclear magnetic resonance by F. Bloch and E. Purcell has revolutionized the field of structural chemistry. The field of molecular biology has also benefited greatly from NMR technology. In this paper, we review the contributions of this technology to the understanding of the structure and dynamics of proteins, nucleic acids, lipids, and membranes.

I. INTRODUCTION

Less than a decade after Felix Bloch's and Edward Purcell's discovery of Nuclear Magnetic Resonance, two reports appeared, one by Shaw and Elksen[1] describing its use to measure hydration of serum albumin and starch, the other by Jacobson, Anderson, and Arnold[2] examining the hydration of DNA. Although it is doubtful that the experiments actually determined the amount of water bound to protein and DNA, the reports aroused interest in the possibilities of using the new method to answer biological questions. In that sense, they inaugurated the era of biological applications of NMR, a field that now, a quarter of a century later, has its own bibliography numbering in the thousands of entries.

A brief essay cannot do justice to the many types of experiments that have been reported. Yet the time seems ripe to pose the question: What sig-

Oleg Jardetzky is Professor and Director, Stanford Magnetic Resonance Laboratory, Stanford University. Norma Wade-Jardetzky is Research Associate, Stanford Magnetic Resonance Laboratory.

nificant problems in molecular biology have been or can be solved by NMR?

Through the years and to this date, much of the work in the field is still best described as the pure joy of spectroscopy—the fascination with NMR signals from complex structures, focusing on attempts to find new ways of improving resolution, sensitivity, and assignment of the hundreds of distinct resonances seen in biological objects. These efforts have advanced NMR technology to a very high degree of sophistication, so that the weak signals in scarce biological materials can be detected with ease and their clustered overlapping resonances more or less readily displayed in their individuality. Perhaps more important is that hand in hand with these efforts has gone a search for the significant problems of contemporary biology onto which this wealth of information could be brought to bear. Was NMR belatedly to confirm facts already known? Was it to supplant existing methods in some areas because of its greater convenience or efficiency? Was it to open to investigation new biological questions not accessible by any other means? Not all the answers are clear, but a sufficient number are, so that we can examine the role of NMR in biological research, not from the point of view of new spectroscopic skill that is required, but in the light of its contributions to the generation of new knowledge about biological systems. Emphasis here is therefore placed on those studies in which NMR has provided direct evidence bearing on the solution of important problems in molecular biology.

II. APPLICABILITY OF NMR TO MOLECULAR BIOLOGY: HISTORY, SCOPE, AND LIMITS

The majority of biological problems to which NMR has been successfully applied concern either (a) the structure or (b) the dynamics of the constituents of living systems; (c) the nature of the interactions between them or (d) the rates and mechanisms of the chemical transformations they carry out.

The first successful attempt to observe a biological macromolecule by NMR was made in 1957 by Saunders, Wishnia, and Kirkwood,[3] who reported a 40 MHz spectrum of the enzyme ribonuclease. It was shown in the same year by Jardetzky and Jardetzky[4] that the spectrum accurately reflected the amino acid composition of the protein, but neither resolution nor sensitivity was adequate to obtain any additional information. Also in 1957, Davidson and Gold[5] first pointed out that the role of paramagnetic ions bound to macromolecules could be studied by their effects on water relaxation, and showed that the Fe^{++} in hemoglobin was effectively inaccessible to solvent.

Early spectrometers permitted no more than crude observations on larger molecules. The earliest biological NMR literature therefore focused on three types of problems: (a) the structure and solution conformation of small molecular weight constituents of living systems, notably amino acids,[6] nucleotides,[7] and ATP-metal complexes[8]; (b) the structure of ligand-macromolecule complexes, insofar as it could be inferred from the observation of ligand spectra[9]; and (c) the use of paramagnetic probes for studying binding sites on DNA[10] and the active sites of metal-requiring enzymes.[11] An interesting early determination of the stereospecificity of an enzymatic reaction by NMR was also reported.[12] Still, a few novel insights foreshadowing a host of later findings could be derived even from the primitive observations. Thus, for example, a comparison of the spectra of native and unfolded proteins[13,14] and the study of polystyrene and poly-γ-benzylglutamate[14] led to two important conclusions[15]: (a) the existence of internal motions in proteins and (b) the dependence of the resonance positions in protein spectra on the internal organization of the protein. The latter is a fortunate quirk of nature, without which NMR would have been all but useless for the study of macromolecules.

Serious NMR spectroscopy of biological structures became possible in 1963, with the introduction of time-averaging for the collection of spectra by Jardetzky, Wade, and Fischer[16] and Klein and Barton.[17] The CW method used initially (CAT NMR) became obsolete by 1969 with the advent of Fourier transform NMR spectrometers largely through the efforts of R. R. Ernst.[18] Nevertheless, it remained in use long enough to allow the completion of several key studies. Progress was also greatly aided by the development of high field superconducting magnets. The first Varian instrument operating at 220 MHz was used for the investigation of proteins by W. D. Phillips and C. C. McDonald at DuPont in 1965,[19] and there has been a steady climb to higher fields since—to 270 MHz at Oxford in 1971, 300 MHz at Varian in 1972, 360 MHz at Stanford in 1974, 470 MHz at Oxford in 1978, and 600 MHz at Carnegie-Mellon in 1979. In addition to the general advances of chemical NMR instrumentation, many other innovations of specific interest to biological work have contributed to the rapid growth of the field. To cite but a few: biosynthetic selective isotopic labeling for the simplification of protein spectra[20]; natural abundance ^{13}C spectroscopy of proteins[21,22]; incorporation of ^{19}F into proteins[23,24]; the use of paramagnetic shift reagents for assignments and determination of distances[25,26]; the use of spin labels,[27,28] and photochemically induced dynamic nuclear polarization (CIDNP)[29,30] for the same purposes; deconvolution techniques[31,32]; two-dimensional Fourier transform spectroscopy (2DFT),[33,34] and solid state high resolution NMR spectroscopy by magic angle spinning[35,36] for resolution enhancement.

New knowledge, which has arisen from the beginnings summarized above, is most easily discussed under the separate headings of the major constituents of living systems. Each of these presents a distinct set of methodological problems to the NMR spectroscopist when questions of their structure, dynamics, interactions, and function are raised: (1) biologically active small molecules, (2) proteins, (3) nucleic acids, (4) water, and (5) lipids and membranes or membrane models.

The extensive and valuable NMR literature on the structure, conformation, and dynamics of small molecules falls by convention more into the realm of chemistry than that of molecular biology, and will not be discussed. We will also omit the literature on water structure in the presence of polymers and in living tissues, because a resolution of the fundamental theoretical difficulties in the interpretation of experimental findings is not in sight.

Two new biological applications of NMR do not properly fall in our chosen domain of molecular biology, but deserve mention for their promise in physiology: NMR studies of metabolism in live tissues, pioneered by the Oxford group,[37] and Lauterbur's zeugmatography-NMR imaging of tissues and organs.[38]

III. STRUCTURE AND DYNAMICS OF PROTEINS

Problems in protein chemistry that have been successfully studied by NMR can be grouped into five categories: A. solution structure; B. behavior of individual residues; C. structure of binding sites; D. dynamics and internal motion; and E. conformational transitions and problems of protein folding.

A. Solution structure of proteins

Two questions are fundamental to our study of protein function: (1) Is the structure of a protein the same in solution as it is in the crystal? and (2) Is the structure of an isolated protein the same as it is when the protein is part of a functioning macromolecular assembly?

A wealth of circumstantial evidence, chemical and spectroscopic, supports the generally accepted affirmative answer to the first question. Yet there are observations that cast doubts on the universal validity of this answer: for example, the occasional ready accessibility to chemical reagents of residues located in the interior of a protein in the crystal structure. Much less is known about the second question. In the one case, of tobacco mosaic virus, where the structure of both the assembled virus and its building block, a double disc, has been determined, both significant similarities and significant differences are apparent.[39] The differences may well be due to the fact that viral RNA is present in the first and not in the second. But in general, the answer is unknown.

NMR is a method that permits a critical test and a definitive answer to both questions.

The basic reason for this is that the NMR spectrum of a protein reflects to only a minor extent its primary structure, i.e., the specific amino acid sequence, and to a large extent its folded (secondary and tertiary) structure.[40] The NMR spectrum of the unfolded protein can be closely approximated by the sum of the spectra of the constituent amino acids, whereas in the folded protein the resonances of chemically equivalent but positionally nonequivalent amino acids are shifted with respect to each other.

If a solution spectrum of a protein—containing dozens of individually resolved lines—is compared to an equally well-resolved solid state spectrum and the two are found to be identical, it can be argued that the protein structures in the two phases are identical. Conversely, spectroscopic differences can be taken to reflect differences in structure or dynamics or both. The probability that identical spectra in solution and in the solid could arise merely by accident is, roughly speaking, inversely proportional to the number of well-resolved lines in the spectra: The less resolved the spectrum, the less well-founded the conclusion.

Our first attempt to make such a comparison is shown in figure 1 for the enzyme lysozyme. The similarity of the two pairs of ^{13}C spectra is striking, but the resolution of the solid state spectrum (~ 7 Hz) is adequate only for a comparison of gross features. The nature of the discrepancies remains to be established, yet they are seen at once not to be large. The answer to the question that has been debated by protein chemists since the publication of the first protein structure (are the solution and the crystal structure identical?) therefore seems to be: almost. Similar experiments can be done with proteins in organized assemblies, except that the background arising from proteins other than the unit of interest would have to be blanked out by selective isotopic substitution.[20,40]

NMR has also provided an abundance of less direct evidence leading to the same conclusion. Essentially, four types of experiments have contributed to it: (1) The structure of substrate and inhibitor binding sites, insofar as it can be inferred from NMR data, is consistent with that found by x ray diffraction. Conversely, the assumption that the structure of an enzyme-inhibitor complex as determined by x ray diffraction also exists in solution accounts for the observed NMR parameters of the complex. This type of study is more fully discussed below (section III,C) and provides strong evidence that *at least* the binding sites and their immediate vicinity are the same in the two phases. (2) Anomalous chemical shifts of methyl groups in the solution NMR spectra of proteins, e.g., such as are observed in lysozyme and hemoglobin, can be explained by the proximity of the observed methyls to aromatic rings, assuming the known crystal structure of the protein. Very good agreement between the observed chemical shift and that calculated from ring anisotropy shielding has been obtained for lysozyme by McDonald and Phillips[41] and for several other proteins by others. (3) Experi-

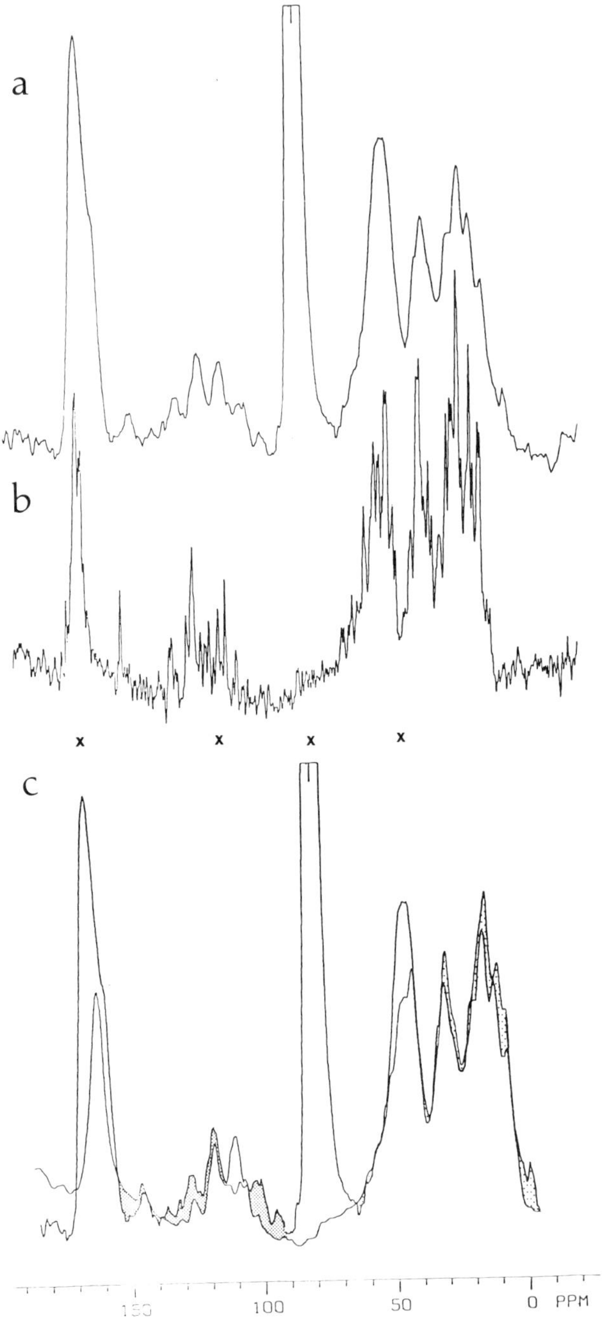

FIG. 1. ^{13}C NMR SPECTRA OF LYSOZYME AT 35.2 MHZ. (a) Crystal powder spectrum with magic angle spinning, 25 °C. (b) Aqueous solution (250 mg/ml, pH 7) spectrum, 25 °C. (c) Aqueous solution spectrum with 200 Hz line broadening superimposed on solid state spectrum, 25 °C. Dotted: difference spectrum. Dense dots: excess area in solid state spectrum. Sparse dots: excess area in solution spectrum. X: rotor peak (center) and side bands.

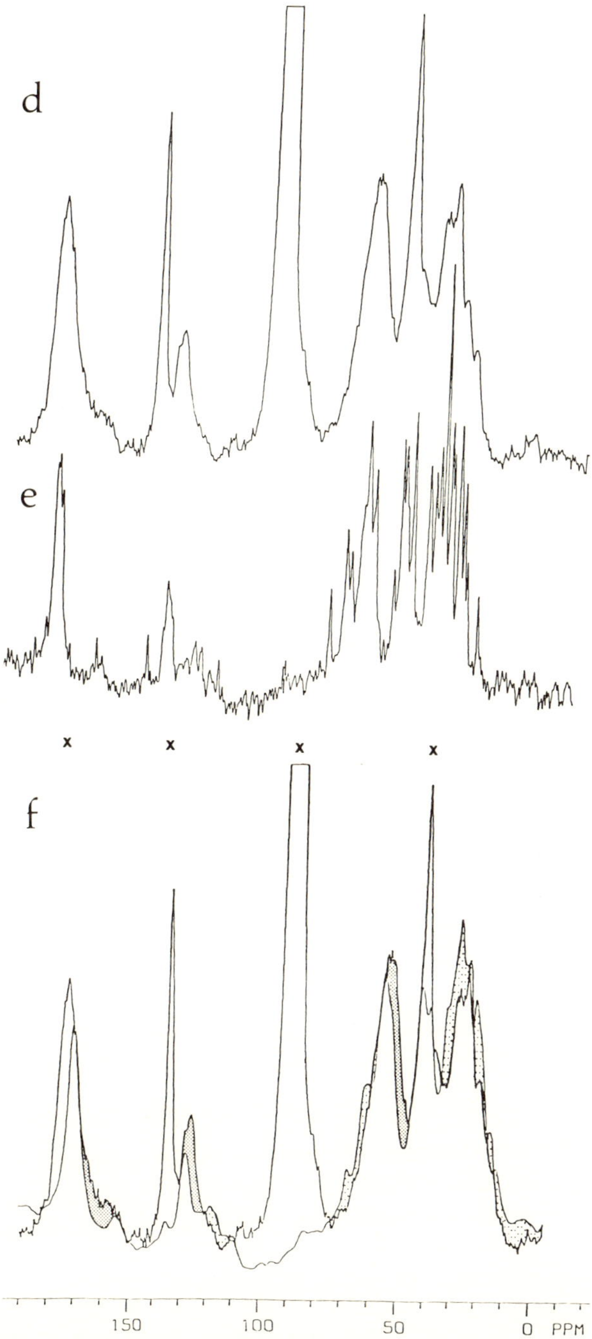

FIG. 1. (d) Crystal powder spectrum with magic angle spinning, 70 °C. (e) Aqueous solution (250 mg/ml, pH 7) spectrum, 70 °C. (f) Aqueous solution spectrum with 200 Hz line broadening superimposed on solid state spectrum, 70 °C. Dotted: difference spectrum. Dense dots: excess area in solid state spectrum. Sparse dots: excess area in solution spectrum. X: rotor peak (center) and sidebands.

ments on spin labeled proteins, such as those of Morrisett, Wien, and McConnell on lysozyme[28] and Dwek et al. on myeloma proteins,[42] yield distances between the spin label at a known site and a proton at another known site (e.g., His 15 and Trp 123 in lysozyme) that are in good agreement with known crystallographic distances. (4) Perturbation of nuclear resonances in proteins by paramagnetic ions bound at known sites sometimes permits the calculation of a distance between the ion and the perturbed group. In a pioneering study, Navon, Shulman, Wyluda, and Yamane[43] calculated the distances in an inhibitor-Mn^{++}-carboxypeptidase complex. Many similar calculations for other enzymes are now recorded in the literature. A bold proposal for the complete quantitative determination of the structure of lysozyme in solution, using lanthanide ions, has been made by R. J. P. Williams and coworkers,[32] and the initial reports suggested that the distances between the lanthanide ion and the amino acid residues in its immediate vicinity were the same in the crystal and in solution. However, serious problems exist in the theoretical interpretation of perturbations of protein resonances by paramagnetic ions.[44,45] Not all the distances calculated by this method are in agreement with those obtained from crystal structure determinations.[46] It would be hazardous to assert on this basis that the solution and crystal structures are different. The discrepancies tend to increase with the distance between the ion and the perturbed group. This reflects not only the increasing error but also the increasing role of paramagnetic ions bound at secondary binding sites, which most proteins have in abundance. If this approach were to succeed completely, it would succeed for a metalloprotein with a unique strong metal binding site and no paramagnetic ions in solution. So far, where the condition of a dominant binding site is even approximately satisfied, the calculated short-range distances are fairly consistent with those seen in the crystal.

The question of the accessibility of exchangeable protons in proteins to solvent is a special aspect of the problem of the solution structure. In a rigid structure or in hydrogen-bonded segments, NH and OH protons in the interior would be essentially inaccessible to solvent; exchangeability would reflect exposure to solvent, although the exchange rates are also known to depend on the character of the X-H bonds involved. Slow exchange of some protons in proteins with solvent protons has been known since the classical isotope experiments of Linderstrøm-Lang in the early 1950s, but NMR has proven to be a particularly easy and convenient method for the study of this question. Glickson, McDonald, and Phillips[47] were the first to report this type of experiment on the NH resonances of tryptophan in lysozyme, and exchangeable protons have since been studied in ribonuclease, chymotrypsin, heme proteins, azurin, and others.[48,49] Assignments of specific Trp and imidazole NH resonances have been made in lysozyme, ribonuclease, and azurin. The essential conclusion emerging from such studies on proteins whose crystal structure is known is that there are marked similarities between the structures in solution and in the crystal. In the case of the copper protein azurin, a cage formed by the peptide segment Tyr 108-His 117 around the Cu^{++} ion making the ion inaccessible to solvent has been proposed by Ugurbil and Bersohn[49]; it remains to be established whether the proposal, based entirely on NMR data, correctly predicts part of the crystal structure.

B. Behavior of individual residues

A unique feature of high resolution NMR of proteins is that it permits the observation of resonances from individual amino acid residues and thus the study of the behavior of these residues for many different purposes. Other spectroscopic methods will occasionally allow the observation of individual residues—for example, in a protein containing a single tryptophan, the specific tryptophan residue and its environment can be studied by fluorescence. But in general, spectra of positionally different, chemically identical residues cannot be resolved, and non-aromatic residues cannot be seen at all, as they can be by NMR. This, in principle, will permit the identification of an individual residue—or constellation of residues—as, for example, the trigger of a functionally significant conformational change, much in the same way as it has permitted the identification of specific residues as part of a binding site (see section III,C). Although no specific instance of such an identification has been reported, it doubtless will be in the near future. At the same time, the feasibility of simultaneously observing resonances from many residues opens the possibilities of a very detailed description of binding interactions and structural transitions. The mechanisms of protein function that will emerge from such descriptions rank for the present among the major unsolved problems of molecular biology.

Although this ultimate use of the observation of resonances from individual amino acid residues in proteins has not yet been achieved for a variety of technical reasons, several more modest but unique contributions to the understanding of protein structure have resulted from it. Among these, one can name (1) identification of titrating groups; (2) determination of microscopic pK values; (3) identification of points of contact; and (4) identification of N- and C-terminal residues.

The prototype study of individual resonances was reported in 1967 by Meadows, Markley, Cohen, and Jardetzky,[50] following up on a report by Bradbury and Scheraga,[51] and involved the determination of the microscopic pK values of the four histidine residues of pancreatic ribonuclease A. Since all four histidines could be assigned by chemical modification and isotopic replacement, identification of individual pK values with specific residues in the sequence became possible. Extensive subsequent work (cited in references 52 and 53) has shown that the assignments were valid for the

EDTA-containing enzyme used in the original study, but otherwise the assignments of the active site His 12 and His 119 were to be reversed.[52,53] This underscored the extraordinary sensitivity of NMR line positions in proteins to the presence of ligands, but the usefulness of the method for the examination of individual resonances remained established. Many studies of histidine residues in proteins have been reported since.

Individual tyrosine resonances have been resolved and at least partially assigned in staphylococcal nuclease,[54] ribonuclease, lysozyme,[45] bovine pancreatic trypsin inhibitor,[55] the N-terminal peptide of the *lac*-repressor, ferricytochrome C and several other proteins. Titration of tyrosine residues is frequently associated with the denaturation of the protein, since the tyrosine pK values fall in the pH range of 10-11. For this reason, microscopic pK values of tyrosines are of lesser interest for studies of enzyme mechanisms than those of the histidines. Tyrosine residues have, on the other hand, attracted attention as frequent constituents of binding sites and indicators of internal mobility, and more will be said about them in these contexts.

Individual tryptophan resonances in the ^{1}H spectra of lysozyme have been assigned by Campbell et al.,[46] and the corresponding resonances in the ^{13}C spectra by Allerhand and his colleagues.[56] The most interesting result of such assignments was the detection of an interaction between Trp 180 and Glu 35, with the spatial relationship between the two residues being altered by the ionization of the carboxyl of Glu 35. The proximity of two tryptophan residues in dihydrofolate reductase had been demonstrated by ^{19}F NMR before the crystal structure of the enzyme became known.[57]

A study of the ^{13}C resonances of the carboxyl group of Glu 81 and the guanido group of Arg 75 in the muscle calcium binding protein parvalbumin led to the conclusion that the two groups formed a salt bridge in the native protein structure.[58]

Identification of N- and C-terminal residues by titration of their NMR resonances has also been repeatedly reported.[40]

The study of individual residues is often difficult, especially in proteins of larger molecular weight, because of the extensive overlap of spectral lines. The difficulty can be largely overcome by a combination of selective isotopic substitution and the use of high frequency spectrometers. Experiments of this type were first done by complete biosynthetic deuteration of the protein, except for the specific amino acid residues to be observed by ^{1}H NMR. The protonated residues were leucine and the protein phycocyanin in the work of Crespi, Rosenberg, and Katz,[59] and tyrosine and tryptophan in staphylococcal nuclease in the study by Markley, Putter, and Jardetzky.[54,60] The power of this approach for simplifying the NMR spectra of macromolecules was clearly established in both studies. In addition, the fortunate choice of tyrosine as the selectively protonated amino acid in our study provided detailed information on the structure of the binding site in staphylococcal nuclease and on the probable folding pathway of the enzyme. The approach was subsequently extended to the selective ^{13}C-^{2}H labeling of histidines in tryptophan synthetase and aspartic transcarbamylase,[61] and to the biosynthetic ^{15}N labeling of hemoglobin.[62]

The full potential of NMR in conjunction with selective isotopic labeling for the study of biological macromolecules has not yet been realized, mainly for reasons of cost. Yet there is little doubt that the combination of these techniques permits the solution of many structural and dynamic problems in molecular biology not otherwise accessible to experimental study.

C. Mapping of binding sites

NMR is a powerful method for the determination of the structure of ligand binding sites in enzymes and other proteins. This is a natural consequence of the fact that individual resonances can be observed for different groups on both the ligand and the protein, and that the spectral parameters of these resonances are very sensitive to interactional effects. Nearly half of the entire literature on protein NMR published in the past decade addresses itself to this problem in a wide variety of proteins.

The first NMR study of the structure of binding sites observing ^{1}H resonances of both the protein and the ligand was that of Meadows, Roberts, and Jardetzky in 1969[63] on the active site of pancreatic ribonuclease A. In this study, information on the geometry of the active site, available from x ray crystallography, was combined with information on chemical shift changes in the spectra of the active site histidines and of the three inhibitors, 2′CMP, 3′CMP, and 5′CMP, to deduce the structure of the enzyme-inhibitor complexes. Differences in the conformations of the inhibitors at the active site were inferred from spectroscopic data, providing an example of "induced fit" in the inhibitor. The consistency of the observed spectroscopic changes with those predicted from the crystal structure of the complex also led to the conclusion that the structures of the active site were identical in the crystal and in solution.

The extensive use of crystallographic information in that study, as well as in many other studies of proteins by NMR (cited in reference 64), has tended to obscure the important fact that interpretation of NMR data in terms of the structure of protein-ligand complexes does not depend on the knowledge of the crystal structure. It is merely simplified by it. The internal consistency of the spectral changes on a series of inhibitors bound to the same site is often sufficient to define the points of contact. This was first demonstrated in the study of the binding of the inhibitor 3′5′ thymidine diphosphate (pdTp) to selectively deuterated staphylococcal nuclease by Markley and Jardetzky.[65]

Exact distances, on the other hand, are not readily obtained from

NMR experiments on diamagnetic species, except for groups strongly shielded by aromatic rings. In the study of ligand-protein complexes, this is a much less severe handicap than is generally recognized, because a large number of constraints is already given by the van der Waals radii of atoms. The correct geometry can often be found by model building from the knowledge of the matrix of pairwise interactions and the assumption of van der Waals contacts. In the case of the pdTp-staphylococcal nuclease complex, the qualitative arguments alone defined a sufficient number of constraints on the possible structure of the enzyme-inhibitor complex to allow only minor deviations of the solution structure from that found in the crystal. A different approach to the study of the same complex was taken by Furie et al.[66] Taking advantage of the fact that the active site of the enzyme contains a calcium ion, and substituting the lanthanide Gd(III) for it, they mapped out the distances from Gd to the pdTp protons in the complex by NMR relaxation measurements. The distances calculated from NMR were consistent with those determined by x ray diffraction. The structure of the complex is shown in figure 2.

An elegant use of NMR in combination with other spectroscopic techniques to define an antibody binding site was recently reported by Dwek et al.[42] High resolution NMR, paramagnetic probes, fluorescence transfer, chemical modification, and the assumption of the invariance of the immunoglobulin fold yielded a structure for the DNP-antibody complex that has not yet been determined by x ray diffraction. If solution and solid state structures of proteins prove generally similar, at least a partial prediction of a protein crystal structure from spectroscopic measurements should be possible. This study may provide a prime example.

Many other examples of the identification of individual residues and ligand groups in enzyme-inhibitor complexes by NMR could be cited, but detailed NMR studies mapping such sites are still relatively few.

In contrast, the use of paramagnetic centers to probe binding sites of enzymes by NMR enjoys continued popularity. It is difficult to assess what lasting contribution to molecular biology will ultimately result from this approach. Its two principal pitfalls, the multiplicity of ion binding sites in most cases, and the use of unverifiable assumptions in the theoretical analysis,[44,45] are well stated by Morris and Dwek[67]: "If there is to be a moral to this article, it is that the perturbations caused by paramagnetic ions on NMR parameters may be severe, and therefore easily measured. While measurement is easy, the interpretation is often difficult."

The reverse might be said of NMR studies on unperturbed diamagnetic protein-ligand systems.

D. Internal motions and protein dynamics

High Resolution NMR is the only method available to the molecular

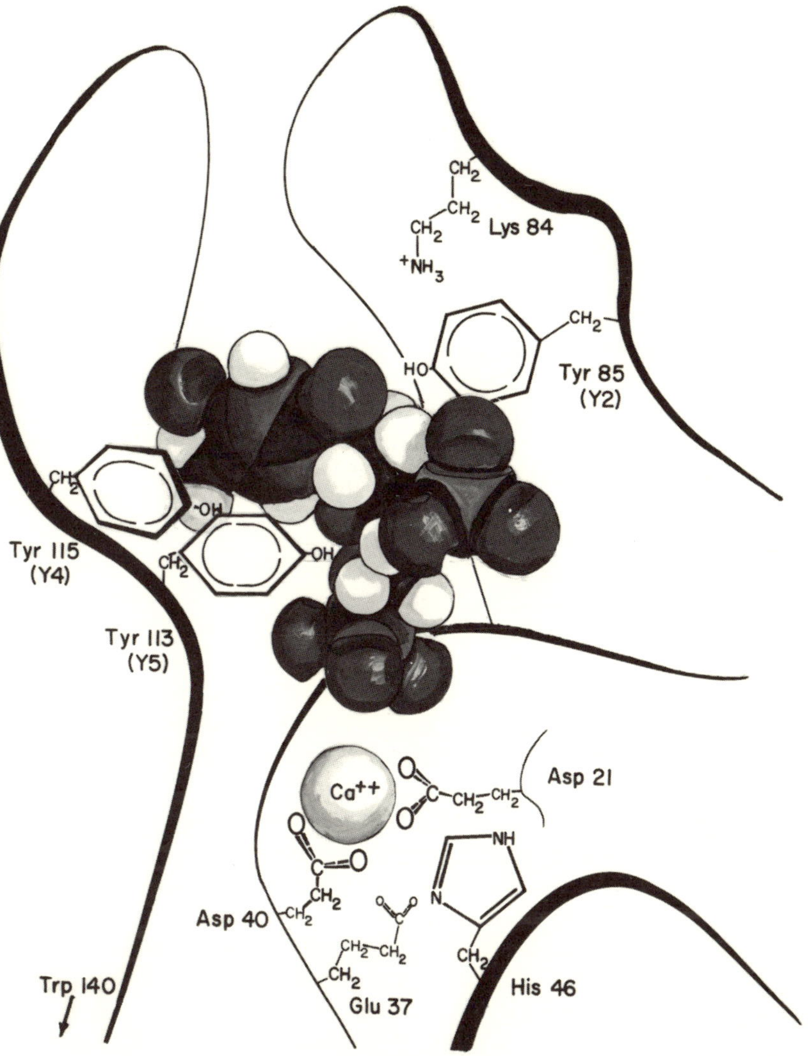

FIG. 2. STRUCTURE OF THE 3′,5′-pdTP-NUCLEASE COMPLEX, showing proximity relationships, which can be inferred from magnetic resonance data.[65,66]

biologist capable of providing detailed, accurate information on macromolecular dynamics. Very little is known about the internal dynamics of proteins, apart from the fact that structural rearrangements are important in their function. The motions by which such rearrangements occur, their extent, and their time scales are for the most part unexplored. Several spectroscopic methods—notably nanosecond fluorescence and depolarized light scattering—have provided evidence for the existence of rapid internal motions and regions of varying flexibility in proteins. Yet the serious drawback of such methods is that they rest on the observation of only a few isolated chromophores and do not permit the reconstruction of a detailed picture of the internal dynamics of a macromolecule. The combination of the feasibility of observing a multitude of individual resonances and the dependence of NMR spectral parameters on the frequency and amplitude parameters of molecular motion makes NMR a method uniquely suitable for this purpose.

It cannot be said at the time of this writing that a molecular dynamic problem of interest in molecular biology has been completely solved by NMR—or by any other technique. This would require not merely the detection of internal motions, but a detailed understanding of their interrelationships and functional significance. Nevertheless, several fundamental developments have taken place that make the success of such an undertaking highly probable in the near future.

Three types of observations pertinent to protein dynamics have been made by NMR: (1) rotation of aromatic rings in the interior of well-defined protein structures; (2) a wide range of rates for amino acid side-chain rotations in different proteins; and (3) occurrence of unusual segmental mobility in some proteins.

The rotation of aromatic side-chains in protein structures is apparent from two types of NMR measurements: (a) averaging of chemical shifts and (b) deviation of the measured relaxation parameters from those predicted from a rigid sphere model.

In several proteins, aromatic resonances do not show a complex splitting pattern, which would be expected in the absence of internal rotation, but reflect motional averaging of chemical shifts. A detailed study of rotational averaging of tyrosine chemical shifts in the bovine pancreatic trypsin inhibitor has been reported.[55,68] Observations in ribonuclease, lysozyme, parvalbumins, and alkaline phosphatase[24] also indicate that ring rotation is a common phenomenon. The rate constant for this rotation has been estimated to be $\sim 10^4$ sec^{-1} from the temperature dependence of the multiplet collapse.[68] On the other hand, much faster rotational motions ($\sim 10^8$ sec^{-1}) have been found by ^{13}C relaxation for phenylalanine rings in parvalbumin.[58] Analysis of the relaxation data suggests that these are librations of relatively low amplitude that have a relatively low probability of exceeding 180° (required for chemical shift averaging).

Since the advent of Fourier transform NMR spectrometers, which make relaxation time measurements comparatively easy, a wealth of experimental relaxation data for both ^{1}H and ^{13}C spectra of proteins has been accumulated. Qualitatively, two facts may be considered as established: (1) The overall diffusional rotation of the protein rarely determines the observed NMR parameters for side-chain resonances; (2) there are differences between relaxation rates of different side-chain groups that could be interpreted as differences in mobility. The quantitative interpretation of such measurements, however, represents a formidable theoretical problem, more fully discussed elsewhere.[69]

Most reported data on protein relaxation have been analyzed using simple motional models—a spinning top on a rigid sphere, or an ellipsoid of revolution. Order of magnitude estimates of correlation times can be made using such models, but for a detailed study of protein dynamics, they are too simple to do justice to the problem. A general theoretical formulation suitable for the analysis of NMR relaxation data on macromolecules with internal motions has only recently been developed by King and Jardetzky.[70] The formulation permits a rigorous definition of the limits of the analysis of relaxation data on systems undergoing complex motions, the systematic testing of specific motional models, and, as an alternative, the treatment of relaxation as an eigenvalue problem with semi-empirical eigenfunctions as amplitude parameters and eigenvectors as rate constants.[71] Analysis of NMR relaxation data on the bovine pancreatic trypsin inhibitor indicates that, on the time scale of motions that such data reflect, convergent solutions of the equations are possible. Thus, internal motions can be described in sufficient detail to permit an understanding of their contribution to protein function.[72,73]

Even a qualitative interpretation of differences in relaxation parameters between residues in different segments of a given protein has led to the discovery of segmental motions and regions of unusual flexibility in several proteins. E. M. Bradbury and colleagues have inferred the existence of random coil, globular, and mixed structures containing either a globular N-terminal half and a flexible C-terminal or a globular core with two "sticky" ends for several histones,[74] and have proposed that a local helix → coil transition involving residues 60-102 occurs in histone F2b. Jardetzky et al.[75] have described segmental flexibility in a region of the isolated coat protein of tobacco mosaic virus, which normally exists as an α-helix (the V-helix) in the assembled virus. Since the V-helix separates the viral RNA from the water in the interior of the virus cylinder, the probable functional role of this local helix → coil transition is to facilitate the embedding of the RNA in the interior of the protein in the course of virus assembly. Myosin (M.W. 450,000) and its high molecular weight fragments were found to contain highly flexible segments.[76] The DNA-binding N-terminal "headpiece" of the *lac*-repressor protein has been shown by NMR to be a distinct, highly mobile domain in the protein structure.[77]

A report on the use of NMR for the study of protein dynamics ten years hence may well contain some of the method's most significant contributions to molecular biology: The fact that internal motions occur in proteins may have been obvious from NMR data for a long time, but the discovery of the diversity of such motions and the development of adequate experimental and theoretical methods for their systematic study are of a very recent origin.

E. Study of conformational transitions and the problem of protein folding

The detection of conformational changes in proteins and a description of their nature is one of the most important applications of protein NMR. Again, ribonuclease was probably the first protein for which a conformational equilibrium involving His 48 was described.[63] No study has yet appeared describing a protein conformational change in a degree of detail comparable to that found in the mapping of binding sites. The most detailed characterization of a conformational change attempted from NMR data is probably that in the calcium binding muscle protein parvalbumin.[58] When one of the two Ca^{++} ions chelated by the protein is removed from its binding site at Lys 96, the salt bridge between Glu 81 and Arg 75 is broken, and Tyr 2 becomes free to rotate. Both effects presumably result from a slight displacement of the E and F α-helices on the hydrophobic core. The slowness with which studies of this type are appearing in the literature reflects the still considerable difficulty of assigning individual resonances to specific amino acid residues in the amino acid sequence. Constituents of binding sites are usually more reactive and therefore more easily identified by any one of several chemical and physical techniques. The direct observation of individual resonances has nevertheless been used to detect an inhibitor-induced local conformational change involving Trp 62, 63 and Ile 98 in lysozyme,[45] as well as a temperature dependent transition involving Trp 108. Similar observations have been reported on ribonuclease and hemoglobin. Many more studies have used extrinsic nuclear spin or paramagnetic probes to detect conformational changes. Among the first was the work of Huestis and Raftery[23] on ribonuclease, lysozyme, and hemoglobin. Radda and his colleagues[78] have monitored structural changes in phosphorylase *b* by a combination of magnetic resonance and fluorescence methods. A less direct method—assessing solvent accessibility to a paramagnetic ion in the interior protein by the measurement of solvent relaxation—has suggested the existence of a conformational change in concanavalin A.[79] Many other suggestions implicating individual residues in conformational changes in different proteins have been published. More detailed descriptions of conformational transitions on the basis of NMR data can be expected as the techniques for resolving and assigning individual resonances improve.

The principal contribution NMR has made thus far to the thorny problem of protein folding is in the detection of kinetic and equilibrium intermediates or alternative structures, especially in ribonuclease and staphylococcal nuclease. Subtle local pre-denaturation changes in ribonuclease have also been found. The major part of the problem, however, remains unsolved. The usefulness of NMR in its study has recently been discussed by Baldwin.[80]

IV. STRUCTURE AND DYNAMICS OF NUCLEIC ACIDS

Applications of high resolution NMR to the study of nucleic acids have been relatively fewer than applications to proteins, although a very large literature exists dealing with the conformation and solution interactions of mono- and oligonucleotides. Three kinds of problems have been investigated with considerable thoroughness: (1) the dynamics and the unwinding of DNA; (2) intercalation of dye and drug molecules into both DNA and RNA; and (3) the solution structure and dynamics of tRNA.

By a combination of ^{1}H, ^{13}C, and ^{31}P NMR measurements on monodisperse DNA fragments 260 and 140 base pairs long, it has been shown that within the rodlike DNA helix the base planes, the deoxyribose sugar, and the sugar-phosphate backbone all experience large fast internal motions.[81,82] The amplitudes A and correlation times τ for these motions could be calculated on a two-state model. For the base planes, $A = \pm 20°$ and $\tau = 1 \times 10^{-9}$ sec; for position 2′ of the deoxyribose sugar, $A = \pm 20°$ to $\pm 33°$ and $\tau = 1 \times 10^{-9}$ sec; for the P-H vector of the sugar-phosphate backbone, $A = \pm 27°$ and $\tau = 2.2 \times 10^{-9}$ sec. Intercalation of ethidium bromide was found to quench the motions of the two base pairs at the intercalation site.[83]

Studies of the interaction of Gene 5 protein from bacteriophage M13 with DNA fragments by Garssen et al.[84] and Coleman et al.[85] using ^{1}H and ^{31}P NMR have led the former group to propose a mechanism of the unwinding of DNA by the protein. This mechanism—tested in detail on the self-complementary tetranucleotide d(pC-G-C-G) with the hydrogen bonded proton resonances of the Watson-Crick base pairs as a probe—consists of two steps: (1) rapid opening and closing of double helical DNA, and (2) subsequent trapping of the single-stranded species by the protein. Stacking of nucleotide bases with tyrosine residues in the protein has been inferred from upfield shifts in the complex. The second step is rate determining, and the rate constant has been estimated to be approximately 10^{8} sec^{-1} M^{-1}. The proposed mechanism depends on the existence of very rapid, spontaneous fluctuations in the DNA structure, so that the double helix opens and closes many times before the single-stranded tetranucleotide interacts with the protein. The evidence for the existence of such fluctuations in this case is kinetic. In a tetranucleotide, the proposed mechanism may hold. Yet in the only double-stranded nucleic acid structure ($tRNA_1^{phe}$), for which the rate of base-pair opening has been measured by saturation transfer, the rate is much slower, and of the order of 10-100 sec^{-1}.[86] The mechanism found in the short nucleotide sequence may not apply to longer double strands, but the usefulness of the method for the study of DNA unwinding, replication, and repair is clear.

Intercalation of Actinomycin D between nucleotides has been studied by Krugh[87] and others, and the intercalations of both Actinomycin D and ethidium bromide into short self-complementary double-stranded DNA fragments, by Patel.[88] The geometry defined by NMR data in solution has proved to be identical to that found in the earlier crystallographic studies of Sobell. Base specificity of Actinomycin D and ethidium bromide for dC-dG sites, selectivity for a pyrimidine 3′-5′ purine sequence, and an increase in the melting temperature of the helix as a result of intercalation, all found previously by other methods, have been confirmed.

Very extensive NMR work has been done on the solution structure and dynamics of tRNA, following the pioneering study by McDonald, Phillips, and Penswick.[89] The sometimes controversial literature has been summarized by Reid and Hurd[90] and by Kearns.[91] Most of the studies rely on the observation of exchangeable protons in the hydrogen bonds of the helical regions. It has been established by careful area measurements that there are 27 such protons in the tRNA from *E. coli*. They yield a well-resolved ^{1}H NMR spectrum at 360 MHz, since the chemical shift of these lines is strongly sequence-dependent (figure 3). This implies that there are 7 ± 1 hydrogen bonded protons in excess of those expected from the known crystal structure of $tRNA_1^{phe}$. These excess protons have been attributed to tertiary base-pairing interactions. It has also been shown that the seven tertiary structure protons disappear at lower temperatures than the secondary structure, indicating that the melting of $tRNA_1^{phe}$ begins with the destruction of the tertiary structure. The sequence of the melting of the secondary structure, which occurs in steps, was found to be rT helix → h_2 U helix – anticodon helix → acceptor helix. Comparisons between the observed NMR spectra of $tRNA_1^{val}$ and spectra calculated on the assumption that its structure is identical to the crystallographically determined structure $tRNA_1^{phe}$ (allowing for differences in base sequence), suggests that the structures are indeed identical, and that there is a universal pattern in the structure of different tRNA's.

V. STRUCTURE AND PHYSICAL PROPERTIES OF MEMBRANES

Biological membranes are known to consist of proteins and phospholipids, arranged, as the bulk of evidence suggests, in a phospholipid bilayer with proteins embedded in it or traversing it. The general features of this so-called "fluid mosaic" model may be regarded as established, but the organization of proteins and lipids in the plane of the membrane remains in essence unknown. Several key problems of membrane organization and

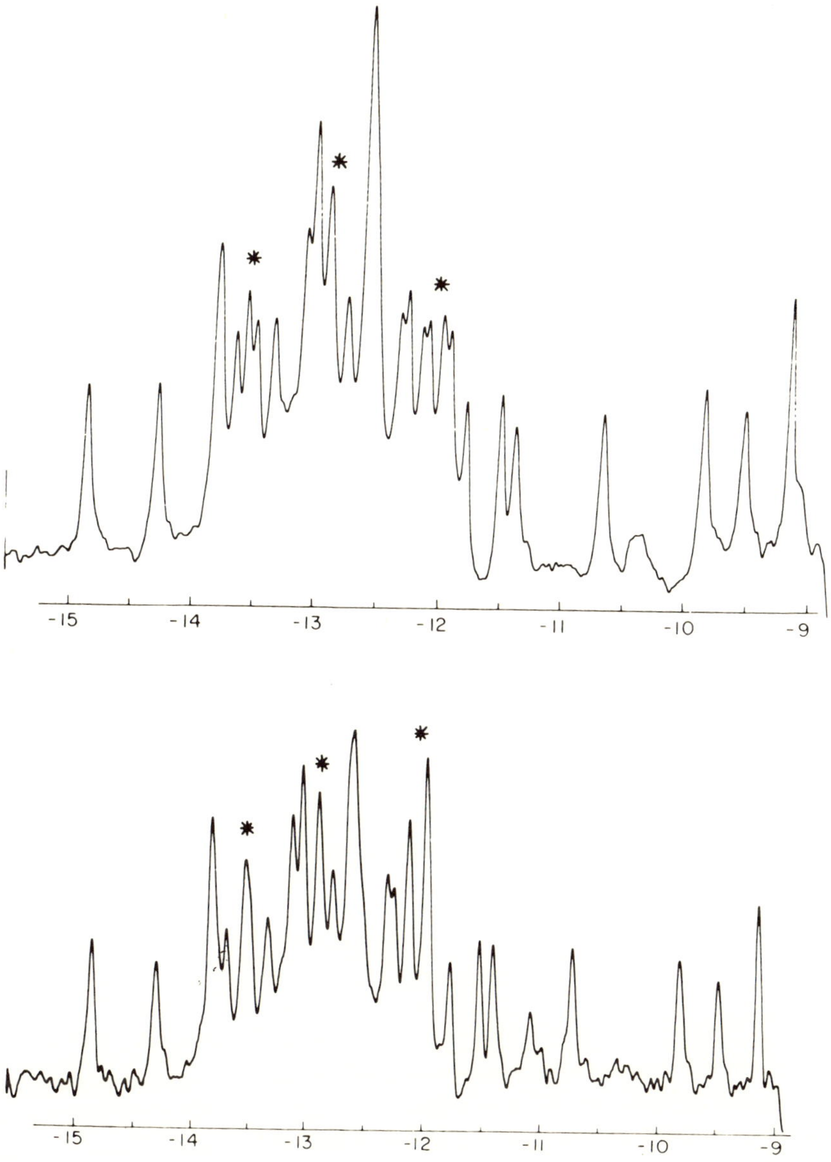

FIG. 3. THE LOW-FIELD 360 MHz NMR SPECTRUM of *E. coli* tRNA$_1^{val}$ under minimal magnesium (upper) and no magnesium (lower) conditions at 35 °C. The tRNA concentration is 1 mM; the no-magnesium solvent contains 10 mM sodium cacodylate, 10 mM EDTA, pH 7.0, and the minimal magnesium spectrum is the same sample supplemented by 2-μL additions of magnesium chloride to a final concentration of approximately 8.8 mM magnesium. The resonances that shift with respect to the excess magnesium spectrum are indicated with asterisks.[90]

dynamics have been investigated—in most cases on model membranes—by ^1H, ^{13}C, ^{31}P, and ^2H NMR. Among the most prominent are: (1) orientation of hydrocarbon chains in phospholipid bilayers; (2) conformation of the polar headgroup of the phospholipids; (3) amplitudes and rates of segmental motions in the hydrocarbon chains; (4) lateral diffusion of phospholipids between the two layers in phospholipid vesicles; (5) rates of transmembrane exchange ("flip flop") of individual phospholipid molecules between the two layers; (6) interaction of small molecules, especially local anesthetics, with lipid bilayers; (7) stabilization of hydrocarbon side chains by proteins, mostly in reconstituted systems; and (8) intermembrane phospholipid exchange.

All of the foregoing structural features and processes can be detected by NMR. At present not all interpretations of experimental findings are clear or consistent, however, and there has probably been more occasion for controversy in this area of biological applications of NMR than in any other. Since our aim is to illustrate the contributions of the method, we will confine ourselves to a few findings where the interpretation is less problematic without attempting a review of the voluminous literature.

Fluidity in membranes was seen by ^1H NMR in the early experiments of Chapman[92] without, however, yielding details of the internal order deduced from the study of spin-labeled membranes by ESR.[93] Differential mobility, increasing from the polar head toward the interior of the bilayer, was first observed by Metcalfe and co-workers[94] using ^{13}C NMR. The dominant effect of incorporating cholesterol into bilayers was found to be a decrease in mobility. The most rigorous NMR study of the ordering of the interior of phospholipid bilayers has been made possible by the use of deuterium resonance on selectively deuterated phospholipids, introduced by Seelig.[95] Deuterium resonance experiments have shown that the order parameter, defined from deuterium quadrupole splittings, remains constant along most of the chain and decreases toward the methyl end. This result differs from that obtained by the use of spin labels, which show a continuous gradient of flexibility along the chain. The difference has been attributed to the perturbation introduced by the bulky nitroxide groups used for spin labeling, since the perturbation is minimal in the case of simple isotopic substitution. The origin of chain flexibility is thought to be in the rapid isomerization between trans and gauche conformations of carbon-carbon bonds.

The structure and orientation of the polar phosphocholine or phosphotidylethanolamine head group has been studied by both deuterium and ^{31}P NMR.[96] The simplest interpretation of the results is that the head group is aligned parallel to the bilayer plane, rapidly oscillating between two conformations in the case of phosphoethanolamine. A different conformation had been proposed on the basis of experiments with lanthanide ions.[97] In criticism, it has been pointed out[95,96] that ionic charge may alter and fix the conformation. In counter-criticism, it could be said that ions are ubiquitous in biological systems, and that the "natural" polar head group conformation in the membrane is more likely to be that of the metal ion complex rather than that of the pure model system. The important conclusion is, probably, that different polar group conformations will be found, depending on the ambient microenvironment.

Asymmetry in the distribution of lipids has been demonstrated in several ways, notably using paramagnetic ions. Since the permeability of phospholipid bilayers and membranes to cations is small and the rate of transmembrane exchange is slow, adding a shift probe to the outside will split the overlapping resonances of the polar head group into two, and adding a paramagnetic relaxation reagent will broaden the resonances of the outer layer. This effect has been demonstrated by ^{1}H, ^{13}C, and ^{31}P NMR, and detailed mapping of membrane sidedness in phospholipid vesicles of different composition has been carried out by Bergelson and Barsukov, who also present an interesting discussion of its possible biological relevance.[98] A combination of magnetic resonance measurements has been used to measure the transmembrane exchange of phospholipids ("flip-flop"), following the classic experiments of Kornberg and McConnell,[99] and values ranging from 0.5-24 hours have been reported for different systems. The significance of the differences is not entirely clear, and the differences may in part reflect artifacts of preparation. But the basic fact that these rates are slow by comparison with the rates of lateral diffusion in the plane of the membrane—estimated as diffusion coefficients $D_L \approx 10^{-7} - 10^{-8}$ cm^2/s—has been established by magnetic resonance methods.

Lau and Chan[100] have observed the facilitation of fusion of lecithin vesicles by the antibiotic alamethicin. The process is related to the phenomenon observed as protein-mediated lipid exchange from one membrane to another.[98] In the protein case, only lipids in the outer layer appear to be exchanged, implying that the transfer protein interacts only with the outer layer. Alamethicin—and possibly other smaller molecules—on the other hand, penetrate both layers.

Many observations on the interaction of ions, drug molecules, and proteins with phospholipid vesicles have been reported, but very few NMR studies have so far progressed to the point of shedding light on the structure of a ligand-phospholipid complex. Lowering of the phase transition temperature seems to be a general feature of increasing the heterogeneity of membrane composition. In many cases, NMR results also indicate decreased mobility in the mixed fluid phase. This suggests that motional rigidity in the fluid phase and the tendency to solidification are not necessarily related in any simple way. The local rigidity, monitored by NMR, may or may not require short-range order, but is compatible with long-range disorder. Solidification, on the other hand, depends on the possibilities of creating long-range order. Comparing the effects of cholesterol and lanosterol on phospholipid vesicles by ^{13}C NMR[101] permits the conclusion that a planar sterol α-face is important for an effective interaction of steroids with hydrocarbon chains, which leads to decreased mobility. Other structural requirements for effective complex formation will undoubtedly be found by similar comparative studies.

The future, however, still holds the answer to the usefulness of NMR in the most difficult area to which it has been applied—that of phospholipid-protein interactions in membranes.

VI. CONCLUSION

There was a time, not so long ago, when writing on NMR in biology consisted of a recital of the physical principles and prophetic praise for its potential in biological research. Barely thirty years after its discovery, and barely fifteen years after the beginning of its serious applications to the study of biological systems, this is no longer necessary. The record of accomplishment, briefly outlined here, is in large measure solid. NMR, viewed as a structural method in molecular biology, illustrates the simple perennial truth that all natural science is ultimately physics and that hence basic discoveries in physics have a profound impact on other fields.

ACKNOWLEDGMENTS

Our work described in this article has been supported by grants from the National Science Foundation (most recently PCM78-07930) and the National Institutes of Health (most recently GM18098 and RR00711), and in part by Merck Sharp and Dohme Research Laboratories, Rahway, New Jersey.

REFERENCES

1. T. M. Shaw and R. H. Elksen, J. Chem. Phys. **21**, 565 (1953).
2. B. Jacobson, W. A. Anderson, and J. T. Arnold, Nature **173**, 772 (1954).
3. M. Saunders, A. Wishnia, and J. G. Kirkwood, J. Am. Chem. Soc. **79**, 3289 (1957).
4. O. Jardetzky and C. D. Jardetzky, J. Am. Chem. Soc. **79**, 5322 (1957).
5. N. Davidson and R. Gold, Biochim. Biophys. Acta **26**, 370 (1957).
6. O. Jardetzky and C. D. Jardetzky, J. Biol. Chem. **233**, 383 (1958).
7. C. D. Jardetzky and O. Jardetzky, J. Am. Chem. Soc. **82**, 222 (1960).
8. M. Cohn, J. Cell. Comp. Physiol. **54**, Suppl. 1, 17 (1959).
9. O. Jardetzky and J. J. Fischer, Proc. Fourth International Conference of Medical Electronics, New York, p. 46 (1961).
10. J. Eisinger, R. G. Shulman, and W. E. Blumberg, Nature **192**, 963 (1961).
11. M. Cohn, Biochemistry **2**, 623 (1963).

12. A. I. Krasna, J. Biol. Chem. **233**, 1010 (1958).
13. M. Saunders and A. Wishnia, Ann. N. Y. Acad. Sci. **70**, 870 (1958).
14. F. A. Bovey, G. V. D. Tiers, and G. J. Filipovitch, J. Polym. Sci. **38**, 73 (1959).
15. O. Jardetzky and C. D. Jardetzky, in *Methods of Biochemical Analysis*, ed. D. Glick, Interscience N.Y. **9**, 235 (1962).
16. O. Jardetzky, N. G. Wade, and J. J. Fischer, Nature **197**, 183 (1963).
17. M. P. Klein and G. W. Barton, Rev. Sci. Instr. **34**, 754 (1963).
18. R. R. Ernst and W. A. Anderson, Rev. Sci. Instr. **37**, 93 (1966).
19. C. C. McDonald and W. D. Phillips, in *Magnetic Resonance in Biological Systems*, ed. A. Ehrenberg, B. G. Malmström, and T. Vängärd (Stockholm, Pergamon Press, 1966), p. 3.
20. O. Jardetzky, Proc. International Conference on Magnetic Resonance, Tokyo, Japan, N-3 (1965).
21. P. C. Lauterbur, Proc. International Conference on Magnetic Resonance in Biological Research (Cagliari, 1969); Appl. Spectrosc. **24**, 450 (1970).
22. A. Allerhand, D. W. Cochran, and D. Doddrell, Proc. Natl. Acad. Sci. U.S. **67**, 1093 (1970).
23. W. H. Huestis and M. A. Raftery, Biochemistry **10**, 1181 (1971).
24. W. E. Hull and B. D. Sykes, J. Mol. Biol. **98**, 121 (1975).
25. C. C. McDonald and W. D. Phillips, Biochem. Biophys. Res. Commun. **35**, 43 (1969).
26. C. D. Barey, A. C. T. North, J. A. Glasel, R. J. P. Williams, and A. V. Xavier, Nature **232**, 236 (1971).
27. G. C. K. Roberts, J. Hannah, and O. Jardetzky, Science **165**, 504 (1969).
28. J. D. Morrisett, R. W. Wien, and H. M. McConnell, Ann. N. Y. Acad. Sci. **222**, 149 (1973).
29. R. Kaptein and S. Oosterhoff, Chem. Phys. Lett. **4**, 195, 214 (1969).
30. R. Kaptein, K. Dijkstra, and K. Nicolay, Nature **274**, 293 (1978).
31. R. R. Ernst, R. Freeman, R. Gestblom, and T. R. Lusebrink, Mol. Phys. **13**, 283 (1971).
32. I. D. Campbell, C. M. Dobson, R. J. P. Williams, and A. V. Xavier, Ann. N. Y. Acad. Sci. **222**, 163 (1973).
33. J. Jeener (1971), quoted in R. R. Ernst, Chimia **29**, 179 (1975).
34. K. Nagayama, K. Wüthrich, P. Bachmann, and R. R. Ernst, Biochem. Biophys. Res. Commun. **78**, 99 (1977).
35. E. R. Andrew, A. Bradbury, and R. G. Eades, Nature **183**, 1802 (1959).
36. M. Cohn, A. Kowalsky, J. S. Leigh, and S. Maričić, in *Magnetic Resonance in Biological Systems*, ed. A. Ehrenberg, B. G. Malmström, and T. Vängärd (Stockholm, Pergamon Press, 1966), p. 45.
37. D. I. Hoult, S. J. W. Busby, D. G. Gadian, R. E. Richards, and P. J. Seeley, Nature **252**, 285 (1974).
38. P. C. Lauterbur, Nature **242**, 190 (1973).
39. A. C. Bloomer, J. N. Champness, G. Bricogne, R. Staden, and A. Klug, Nature **276**, 362 (1978).
40. G. C. K. Roberts and O. Jardetzky, Adv. Prot. Chem. **24**, 447 (1970).
41. C. C. McDonald and W. D. Phillips, J. Am. Chem. Soc. **91**, 1513 (1969).
42. R. A. Dwek, S. Wain-Hobson, S. Dower, P. Gettins, B. Suttón, S. J. Perkins, and D. Givol, Nature **266**, 31 (1977).
43. G. Navon, R. G. Shulman, B. G. Wyluda, and T. Yamane, Proc. Natl. Acad. Sci. U.S. **60**, 86 (1968).
44. D. G. Agresti, R. E. Lenkinski, and J. D. Glickson, Biochem. Biophys. Res. Commun. **76**, 711 (1977).
45. M. Epstein and J. Reuben, Biochim. Biophys. Acta **481**, 164 (1977).
46. I. D. Campbell, C. M. Dobson, and R. J. P. Williams, Proc. R. Soc. London, series A. **345**, 23, 41 (1975).

47. J. D. Glickson, C. C. McDonald, and W. D. Phillips. Biochem. Biophys. Res. Commun. **35**, 492 (1969).
48. D. J. Patel, C. K. Woodward, and F. A. Bovey, Proc. Natl. Acad. Sci. U.S. **69**, 599 (1972).
49. K. Ugurbil and R. Bersohn, Biochemistry **16**, 895 (1977).
50. D. H. Meadows, J. L. Markley, J. S. Cohen, and O. Jardetzky, Proc. Natl. Acad. Sci. U.S. **58**, 1307 (1967).
51. J. H. Bradbury and H. A. Scheraga, J. Am. Chem. Soc. **88**, 4240 (1966).
52. J. L. Markley, Acc. Chem. Res. **8**, 70 (1975).
53. M. Brauer and F. W. Benz, Biochim. Biophys. Acta **533**, 186 (1978).
54. I. Putter, J. L. Markley, and O. Jardetzky, Proc. Natl. Acad. Sci. U.S. **65**, 395 (1970).
55. G. H. Snyder, R. Rowan, S. Karplus, and B. D. Sykes, Biochemistry **14**, 3765 (1975).
56. E. Oldfield, R. S. Norton, and A. Allerhand, J. Biol. Chem. **250**, 6368, 6381 (1975).
57. B. J. Kimber, J. Feeney, G. C. K. Roberts, B. Birdsall, D. V. Griffiths, A. S. V. Burgen, and B. D. Sykes, Nature **271**, 184 (1978).
58. D. J. Nelson, S. J. Opella, and O. Jardetzky, Biochemistry **15**, 552 (1976).
59. H. L. Crespi, R. M. Rosenberg, and J. J. Katz, Science **161**, 795 (1968).
60. J. L. Markley, I. Putter, and O. Jardetzky, Science **161**, 1249 (1968).
61. D. T. Browne, G. L. Kenyon, H. Packer, H. Sternlicht, and D. M. Wilson, J. Am. Chem. Soc. **95**, 1316 (1973).
62. A. Lapidot and C. S. Irving, J. Am. Chem. Soc. **99**, 5488 (1977).
63. D. H. Meadows, G. C. K. Roberts, and O. Jardetzky, J. Mol. Biol. **45**, 491 (1969).
64. K. Wüthrich, *NMR in Biological Research: Peptides and Proteins* (North-Holland, Amsterdam, 1976).
65. J. L. Markley and O. Jardetzky, J. Mol. Biol. **50**, 223 (1970).
66. B. Furie, J. H. Griffin, R. J. Feldman, E. A. Sokoloski, and A. Schechter, Proc. Natl. Acad. Sci. U.S. **71**, 2833 (1974).
67. A. T. Morris and R. A. Dwek, Quart. Rev. Biophys. **10**, 421 (1977).
68. G. Wagner, A. De Marco, and K. Wüthrich, Biophys. Struct. Mech. **2**, 139, 159 (1976).
69. O. Jardetzky, in *NMR in Biochemistry*, Proc. Symposium in honor of M. Cohn (Marcel Dekker, Philadelphia, 1979).
70. R. King and O. Jardetzky, Chem. Phys. Lett. **55**, 15 (1978).
71. R. King, R. Maas, M. Gassner, R. K. Nanda, W. W. Conover, and O. Jardetzky, Biophys. J. **24**, 103 (1978).
72. A. Ribeiro, R. King, C. Restivo, and O. Jardetzky, J. Am. Chem. Soc. **102**, 4040 (1980).
73. A. Ribeiro, N. Wade-Jardetzky, R. King, and O. Jardetzky, Appl. Spectroscopy **34**, 299 (1980).
74. G. E. Chapman, P. G. Hartman, and E. M. Bradbury, Eur. J. Biochem. **61**, 69 (1976).
75. O. Jardetzky, K. Akasaka, D. Vogel, S. Morris, and K. C. Holmes, Nature **273**, 564 (1978).
76. S. Highsmith, K. Akasaka, M. Konrad, R. S. Goody, K. C. Holmes, N. Wade-Jardetzky, and O. Jardetzky, Biochemistry **18**, 4238 (1979).
77. N. G. Wade-Jardetzky, R. P. Bray, W. W. Conover, O. Jardetzky, N. Geisler, and K. Weber, J. Mol. Biol. **128**, 259 (1979).
78. A. Bennick, R. A. Dwek, J. A. Griffiths, and G. K. Radda, Ann. N. Y. Acad. Sci. **222**, 175 (1973).
79. R. D. Brown, C. F. Brewer, and S. H. Koenig, Biochemistry **16**, 3883 (1977).
80. R. L. Baldwin, Ann. Rev. Biochem. **44**, 453 (1975).
81. M. Hogan and O. Jardetzky, Proc. Natl. Acad. Sci. U.S. **76**, 6341 (1979).
82. M. Hogan and O. Jardetzky, Biochemistry **19**, 3460 (1980).
83. M. Hogan and O. Jardetzky, Biochemistry **19**, 2079 (1980).

84. G. J. Garssen, C. W. Hilbers, J. G. G. Schoenmakers, and J. H. van Boom, Eur. J. Biochem. **81**, 453 (1977).
85. J. E. Coleman, R. A. Anderson, R. G. Ratcliffe, and I. M. Armitage, Biochemistry **15**, 5419 (1976).
86. P. D. Johnston and A. G. Redfield, Nucleic Acids Res. **4**, 3599 (1977).
87. T. R. Krugh, E. S. Mooberry, and Y. C. C. Chias, Biochemistry **16**, 740 (1977).
88. D. J. Patel and L. L. Cannel, Proc. Natl. Acad. Sci. U.S. **73**, 3343 (1976).
89. C. C. McDonald, W. D. Phillips, and R. Penswick, Biopolymers **3**, 609 (1965).
90. B. R. Reid and R. E. Hurd, Acc. Chem. Res. **10**, 396 (1977).
91. D. R. Kearns, Ann. Rev. Biophys. Bioeng. **6**, 477 (1977).
92. D. Chapman and S. A. Penkett, Nature **211**, 314 (1966).
93. H. M. McConnell and B. G. McFarland, Ann. N. Y. Acad. Sci. **195**, 207 (1972).
94. J. C. Metcalfe, N. J. M. Birdsall, J. Feeney, A. G. Lee, Y. K. Levine, and P. Partington, Nature **233**, 199 (1971).
95. J. Seelig, Q. Rev. Biophys. **10**, 353 (1977).
96. J. Seelig, Biochim. Biophys. Acta **515**, 105 (1978).
97. H. Hauser, M. C. Phillips, B. A. Levine, and R. J. P. Williams, Nature **261**, 390 (1976).
98. L. D. Bergelson and L. I. Barsukov, Science **197**, 224 (1977).
99. R. D. Kornberg and H. M. McConnell, Proc. Natl. Acad. Sci. U.S. **68**, 2564 (1971).
100. A. L. Y. Lau and S. I. Chan, Proc. Natl. Acad. Sci. U.S. **72**, 2170 (1975).
101. P. L. Yeagle, R. B. Martin, A. K. Lala, H. K. Lin, and K. Bloch, Proc. Natl. Acad. Sci. U.S. **74**, 4924 (1977).

DYNAMICALLY POLARIZED TARGETS: AN OUTGROWTH OF MAGNETIC RESONANCE

by Carson D. Jeffries

ABSTRACT

The use of polarized targets in high-energy elementary particle physics is an important technique for the study of the spin-dependence of the nuclear interactions. In this paper, we trace the development of practical targets and review how the dynamical enhancement of the polarization is based on the principles of magnetic resonance.

When Felix Bloch and Edward Purcell first observed[1,2] the magnetic resonance signals from protons in 1945, they had no idea of the ensuing intense developments, generalizations, and extensions of their ideas. Even less could they have imagined the many applications of NMR: several authors in this volume describe the significance of some of these developments. In my essay I will trace the development of dynamically polarized nuclear targets in which, say, all the protons in quite sizable hydrogenous targets can be oriented at will in any direction, thus affording study of spin dependent interactions in particle scattering. Such targets are now a standard tool of nuclear and elementary particle physics, and are found in dozens of international laboratories. The ideas involved in the development of these targets come directly from the discipline of magnetic resonance; this will be a personal account of that development, including the early contributions of Felix Bloch to the concepts of dynamic nuclear polarization. Aside from the target aspects, dynamic polarization can be used to enhance the magnetic resonance signal-to-noise ratio by many orders.

I arrived at Stanford University early in 1946 to begin graduate work,

---

Carson Jeffries is Professor of Physics at the University of California, Berkeley.

just shortly after Bloch and his colleagues had succeeded in observing NMR signals from protons in water. Being young and insufficiently aware of the real world of physics and the wave of post-war developments, I did not appreciate the significance of the discovery of NMR until a year or so later, when I began doctoral research under Bloch's spirited guidancc. My project was to measure precisely the proton magnetic moment in nuclear magnetons.[3] I learned the mysteries of NMR and anti-cyclotrons, how to find signals hidden in noise, and how to do research on a shoestring. In this Festschrift I want to acknowledge, with love and gratitude, my extreme good fortune in having Felix Bloch as my mentor. Felix was alive with ideas, energy, integrity, marvelous physical insight, and a marked capacity—a passion—for hard and honest work. Having good Swiss personal habits, Felix regularly went home to his family for lunch and a nice nap. He returned full of energy and fresh ideas at around tea time. Long, detailed, and loud discussions at the blackboard and in the laboratory with students and colleagues were regular, and in fact, vital phenomena—essential to the rapid development of the ideas, the experiments, the understanding, and the detailed theory. Momentum was transferred from Felix to his students, both linear and angular, with a sometimes dizzying effect. Felix is very gifted in his ability to take apart an idea or a new experimental result, critically and analytically—in detail—with enthusiastic rapidity. I recall my years at Stanford as some of the most exciting in my life. Finally, and most important for me personally, what I got from Felix was confidence: his belief in me—that I could do a hard experiment independently.

In a magnetic field H at temperature T, the proton polarization from the Zeeman interaction with the magnetic moment μ_n is $p \approx \mu_n H/kT \approx 10^{-4}$ $H(kG)/T(\mathrm{K})$; this is too small to be useful at easily accessible fields and temperatures. The first idea that the nuclear polarization could be dynamically enhanced (by the ratio $\mu_e/\mu_n \approx 10^3$, where μ_e is the electron magnetic moment) came from Overhauser[4] in 1953. In an elegant but not very physically transparent argument he predicted the remarkable result that nuclei of spin **I** and conduction electrons of spin **S** in a metal, with interaction $A(\mathbf{I}\cdot\mathbf{S})$, could be given a polarization enhanced by μ_e/μ_n, by strongly inducing the rf (microwave) resonance of the conduction electron spin resonance; this was experimentally confirmed.[5] The Overhauser effect seemed mysterious and intriguing to me when I came to Berkeley to begin my career, but I didn't really understand it until I heard a talk by Bloch in December 1953 at a Stanford Symposium on magnetic resonance.[6] Already the field of NMR was turning to investigation of the environment of nuclei, measurements of the electric field gradient tensors in crystals, ordering in

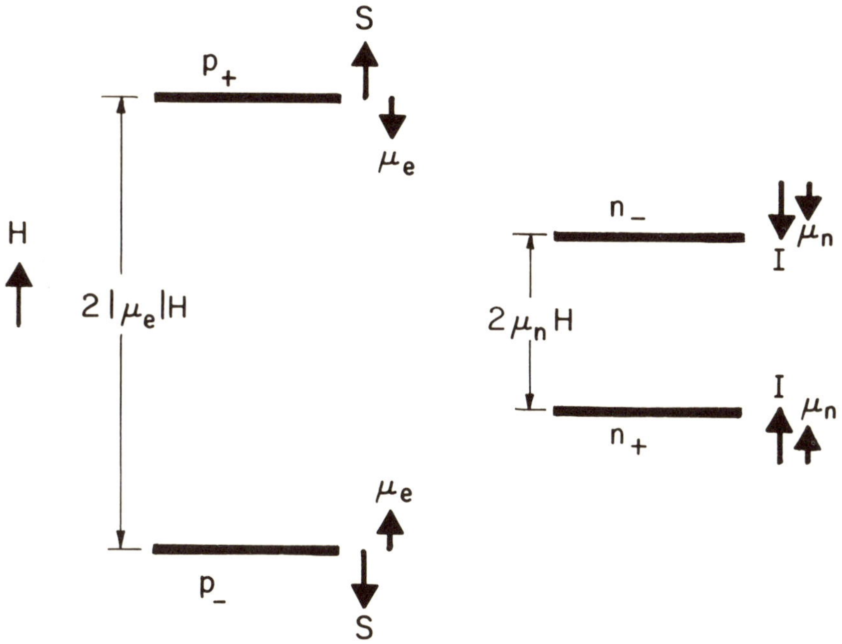

FIG. 1. ZEEMAN ENERGY LEVELS and relative populations of conduction electron **S**; Zeeman energy levels and relative populations of nuclei **I** in the metal.

alloys, electron density in metals, relaxation mechanisms, spin echoes, and other dynamic effects. In particular, Bloch, having quickly understood the generality of Overhauser's idea, presented a simple physical model, which I reproduce here from my notes taken in 1953.

For nuclei of spin **I** and conduction electrons of spin **S** in hyperfine (hfs) interaction in a metal in a field **H** the Hamiltonian is

$$\mathcal{H} = -\boldsymbol{\mu}_e\cdot\mathbf{H} - \boldsymbol{\mu}_n\cdot\mathbf{H} + A(\mathbf{S}\cdot\mathbf{I}) \quad . \tag{1}$$

The conduction electrons move so rapidly, however, that the static part of the hfs term is averaged to zero. Imagine then a system of electrons occupying the two Zeeman states, split by $-2\mu_e H$, with probability p_+ and p_- for spins parallel and anti-parallel to **H**, respectively (we take μ_e negative). Similarly the nuclei occupy the two nuclear Zeeman states, split by $2\mu_n H$, with populations n_+ and n_- for spins parallel and anti-parallel to **H** (we take μ_n positive). This situation is illustrated in figure 1. The electrons are strongly coupled to lattice phonons and quickly relax to the thermal equilibrium ratio

$$\frac{p_+}{p_-} = e^{2\mu_e H/kT} \qquad (2)$$

But the nuclei are essentially uncoupled from the lattice; their relaxation is almost exclusively through the thermal fluctuations in the hfs term $A(t)$ $(S_+I_- + S_-I_+)$. This means that a nucleus is flipped up only by a simultaneously mutual spin flop down with an electron as it collides with the nucleus. The nuclear relaxation rate equation is

$$-\frac{dn_+}{dt} = \frac{dn_-}{dt} = -\frac{1}{T_{1n}}\left[p_+n_- - p_-n_+e^{2(\mu_e-\mu_n)H/kT}\right] \qquad (3)$$

where T_{1n} is the nuclear relaxation time; the p_+n_- term arises from the $A(t)$ S_-I_+ operator; and the p_-n_+ term arises from the $A(t)$ S_+I_- operator and must be weighted by the Boltzmann factor in the final state, since these operators represent thermal processes. In steady state, Eq. (3) can be written

$$\frac{n_+}{n_-} = \frac{p_+}{p_-} e^{-2(\mu_e-\mu_n)H/kT} \quad . \qquad (4)$$

As expected, in thermal equilibrium, from Eq. (2) this yields the nuclear polarization $p \equiv \frac{n_+ - n_-}{n_+ + n_-} = \tanh(\mu_n H/kT) \approx \mu_n H/kT$. Now suppose that the electron spins are not in thermal equilibrium, but that strong rf (microwave) saturation of the electron spin resonance makes $p_+ \approx p_-$, yielding the enhanced ratio $(n_+/n_-)_{\mathrm{enh}} \approx \exp(-2\mu_e H/kT)$ and the enhanced nuclear polarization

$$p_{\mathrm{enh}} \approx -\tanh(\mu_e H/kT) \approx -\mu_e H/kT \quad , \qquad (5)$$

larger than Eq. (4) by $(\mu_e/\mu_n) \approx 10^3$. The nuclei are polarized *as if* they had a magnetic moment as large as the electron.

In presenting this simple physical model of Overhauser's result, Bloch emphasized that the enhancement phenomenon is not restricted to metals but could apply to any electron-nuclear spin coupled system characterized by strong $S_\pm I_\mp$ flip-flop relaxation; negligible $I_\pm S_\pm$ flip-flip relaxation; negligible direct $I_\pm$ nuclear relaxation; and the ability to saturate the electron spin resonance by an external rf field. Others,[7,8,9,10] including Overhauser himself,[11] also gave simple kinetic or thermodynamic arguments to explain the effect. For me personally the most influential and pivotal paper was that of Abragam,[12] who analyzed dynamic nuclear polarization for three cases:

1) nuclei in liquids containing paramagnetic impurities; 2) nuclei in hfs coupling to paramagnetic ions in crystals; and 3) nuclei in long range dipole-dipole coupling to dilute paramagnetic impurities in solids. He concluded that in cases 1) and 2), generalized Overhauser enhancements could be achieved, but that none was to be expected in case 3) because the flip-flop and flip-flip relaxations were equally likely, negating any net nuclear enhancement.

The case of a paramagnetic ion in a crystal can be simply treated assuming a Hamiltonian

$$\mathcal{H} = \beta \mathbf{S}\cdot\mathbf{g}\cdot\mathbf{H} + \mathbf{S}\cdot\mathbf{A}\cdot\mathbf{I} \quad . \tag{6}$$

The states $|M,m\rangle$ of this system are shown in the level diagram of figure 2, where $M = \langle S_z \rangle$ and $m = \langle I_z \rangle$, and $I = \frac{1}{2}$ and $S = \frac{1}{2}$ are assumed.

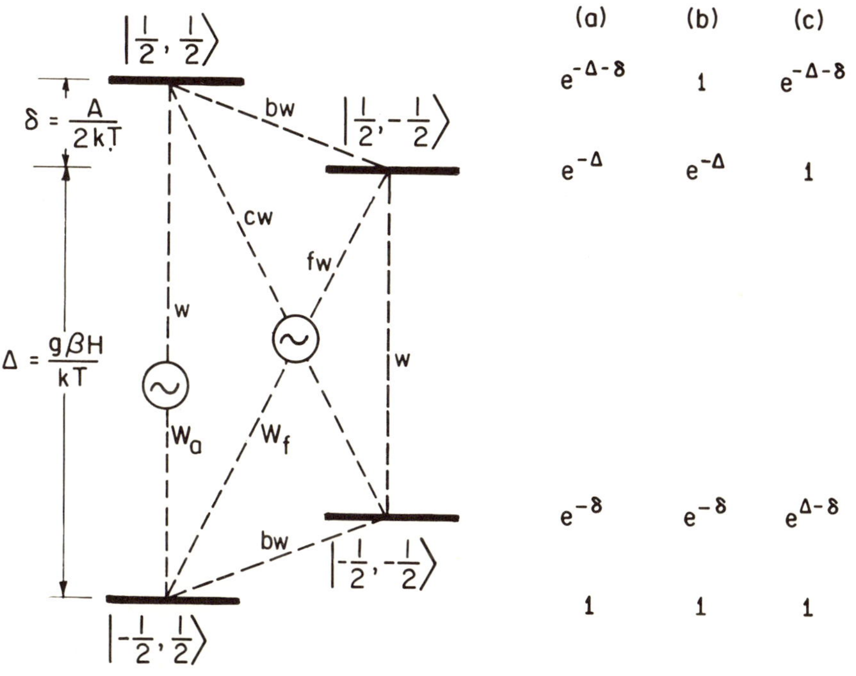

FIG. 2. ENERGY LEVELS AND STATES $|M,m\rangle$ of a paramagnetic ion **S** and its nucleus **I** in a crystal in a magnetic field. The relaxation rates are w, bw, cw, and fw. An applied rf oscillator can induce allowed transitions at the rate W_a, or "forbidden" transitions at the rate W_b. Column (a) shows the thermal equilibrium population. Column (b) shows the dynamic populations resulting if W_a is saturated assuming $f >> c,b$. Column (c) shows the dynamic populations produced if W_f is saturated.

Column (a) shows the relative thermal equilibrium populations $N(M,m)$ of the system; the dominant relaxation rate w arises from the thermal fluctuations of the crystalline electric field. If the allowed transition W_a ($\Delta M = \pm 1, \Delta m = 0$) is saturated, the populations $N(\frac{1}{2},\frac{1}{2})$ and $N(-\frac{1}{2},\frac{1}{2})$ are driven to the same value, say unity, column (b). If we assume, as is often the case, that the other relaxation rates are such that $f >> c,b$, then it is apparent that the other two states take the populations shown in column (b), yielding an enhanced dynamic nuclear polarization $p \approx (\Delta + \delta)/4$. This diagram and example are very similar to the Overhauser effect in metals.

It is noteworthy that Bloch,[13] in a long elegant paper on the "dynamical theory of nuclear induction," had solved the problem of calculating the dynamic populations of a multi-level spin system under the general influence of all possible relaxation transitions and of pumping. He showed that the overall result could be written in a form similar to Kirchoff's laws for currents out of each junction point of a resistive network (figure 3). The resistances are related to reciprocal relaxation rates w^{-1}, the currents to the rf transition rates W, and the electric potential at any point to the population. Bloch called this the "equivalent dc circuit for the (generalized) Overhauser effect." This is a very useful analogy and typical of his fondness for simple classical physical equivalents of seemingly complex microscopic phenomena. Another example is his invention of a non-linear mechanical demonstration of the Josephson effect.

Can the Overhauser effect fail? In Eq. (6) and figure 2, if the hfs term is such that $f \approx c$, then saturation of an allowed transition ($\Delta M = \pm 1$)

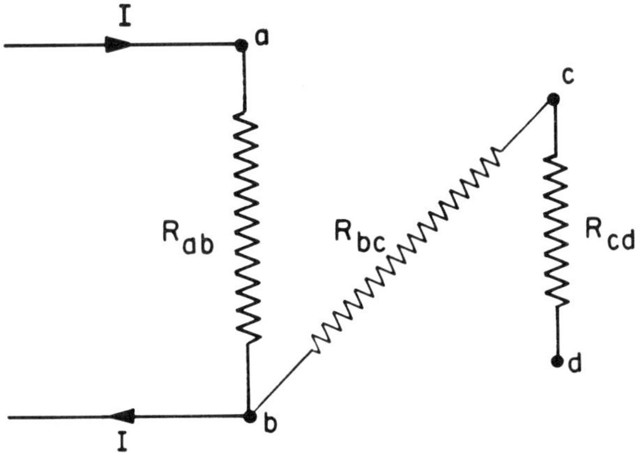

FIG. 3. F. BLOCH'S "equivalent dc circuit for the Overhauser effect," from reference 13. The points a, b, c, d correspond to the states $|\frac{1}{2},\frac{1}{2}\rangle$, $|-\frac{1}{2},\frac{1}{2}\rangle$, $|\frac{1}{2},-\frac{1}{2}\rangle$ and $|-\frac{1}{2},-\frac{1}{2}\rangle$, respectively, in figure 2. Current I corresponds to saturation of W_a in figure 2. This circuit solves the problem of column (b), figure 2.

does not produce dynamic polarization, essentially because the relaxation rates fw and cw "short circuit" each other. This can be seen from Bloch's circuit (figure 3) by adding a resistor $R_{ad} = R_{bc}$; this would maintain the potentials, and the populations, equal at points c and d, giving no polarization.

The next advance came with my suggestion[14] that even if $f=c$, rf saturation of a "forbidden" transition $W_f(\Delta M = \pm 1, \Delta m = \mp 1)$ can indeed induce the large dynamic populations shown in figure 2, col. (c), and a large dynamic polarization $p \approx -\Delta/2$. This was experimentally demonstrated[15] by saturation of W_f for Co^{++} ions, where saturation of W_a failed to produce polarization. We note that in both cases nuclear polarization arises from a simultaneous flip-flop of an electron and a nuclear spin: i) rf saturation of forbidden transitions drives this flip-flop directly, and is more immune to unfavorable competing relaxation rates; ii) in saturation of an allowed transition, the method works only if a favorable relaxation rate ($I_{\pm} S_{\mp}$) is dominant.

Although figure 2(c) assumes an S,I coupling of the form $\mathbf{S} \cdot \mathbf{I}$, it occurred to me, and independently to Abragam and Proctor,[16] that saturation of forbidden transitions would also produce a polarization in the case of S,I dipolar coupling, where indeed the flip-flip (S_+I_+) and flip-flop (S_+I_-) relaxation rates are equal, and saturation of the allowed transitions fails.[12] The idea of saturation of forbidden transitions in the dipolar coupling case was the key to the development of polarized nuclear targets, our special concern in this essay.

To fix ideas, consider the substance extensively studied at Berkeley,[17] a single crystal of $La_2Mg_3(NO_3)_{12} \cdot 24H_2O$, in which ~1% of diamagnetic La^{3+} ions are replaced by paramagnetic Nd^{3+} ions (denoted as Nd:LMN). The Nd ions form a $\mathbf{S}_k$ dilute electron spin system, and the protons in the waters of hydration form an $\mathbf{I}_i$ nuclear spin system. Figure 4 shows a typical experimental arrangement to keep in mind: the crystal is mounted in a microwave cavity in a field $H \sim 10^4$ kOe and in a helium bath at $T \sim 1°K$. The crystal is also coupled to an rf coil that measures directly the proton polarization from the NMR signal at ν_n.

The Hamiltonian for this system is

$$\mathcal{H} = \beta \sum_k \mathbf{H} \cdot \mathbf{g} \cdot \mathbf{S}_k - g_n \beta \sum_i \mathbf{H} \cdot \mathbf{I} + \sum_{i,k} \mathcal{H}_{dd}(\mathbf{I}_i, \mathbf{S}_k) + \sum_{i,j} U_{ij} \quad (7)$$

where the terms represent the electron Zeeman, nuclear Zeeman, electron-nuclear dipole, and nuclear-nuclear dipole interactions, respectively. The nuclei greatly outnumber the electron spins, and the levels labeled in zero order by the high field states $|M,m\rangle^0$ of a typical I,S pair are shown in

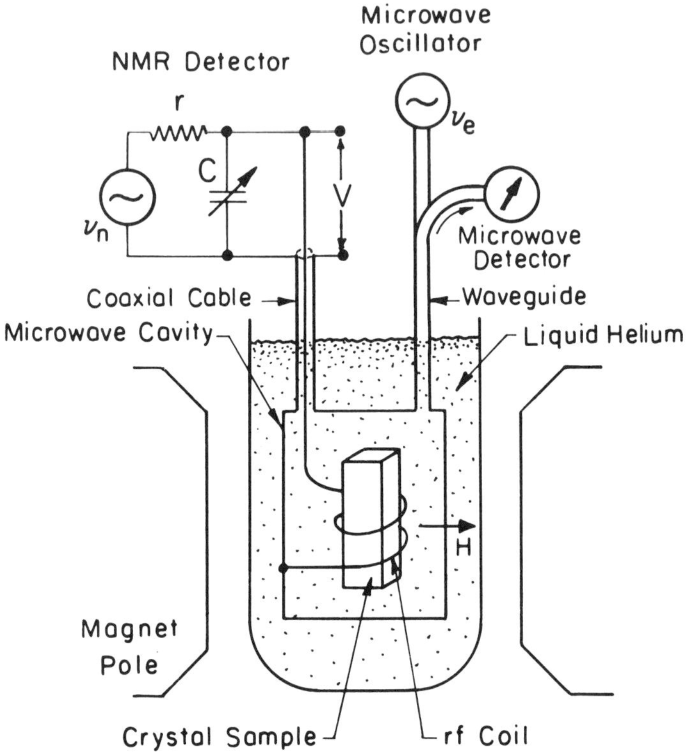

FIG. 4. TYPICAL APPARATUS FOR DYNAMIC NUCLEAR POLARIZATION; see reference 17.

figure 5. Only the Zeeman terms give a significant contribution to the energy spacing. The I_xS_z terms in $\mathcal{H}_{dd}$ admix slightly these states, however, enabling the forbidden transitions W_2 and W_3, weaker than the allowed transitions W_1 by the small factor

$$\sigma \approx \frac{3}{10}\left(\frac{g\beta}{r^3H}\right)^2 , \tag{8}$$

where r is the distance between I and S. The Nd ion, having electric multipole moments, is strongly relaxed by the thermal fluctuations of the crystalline field at a rate w_1. The proton is virtually uncoupled from the lattice; however, the admixing of states that enables W_2 and W_3 also produces flip-flop and flip-flip relaxation rates $w_2 \approx \sigma w_1$ and $w_3 \approx \sigma w_1$: i.e., the proton is relaxed only through mutual spin flips with the Nd ion. Suppose, in figure 5, that W_3 is strongly saturated ($W_3 >> w_3$) by setting the microwave oscil-

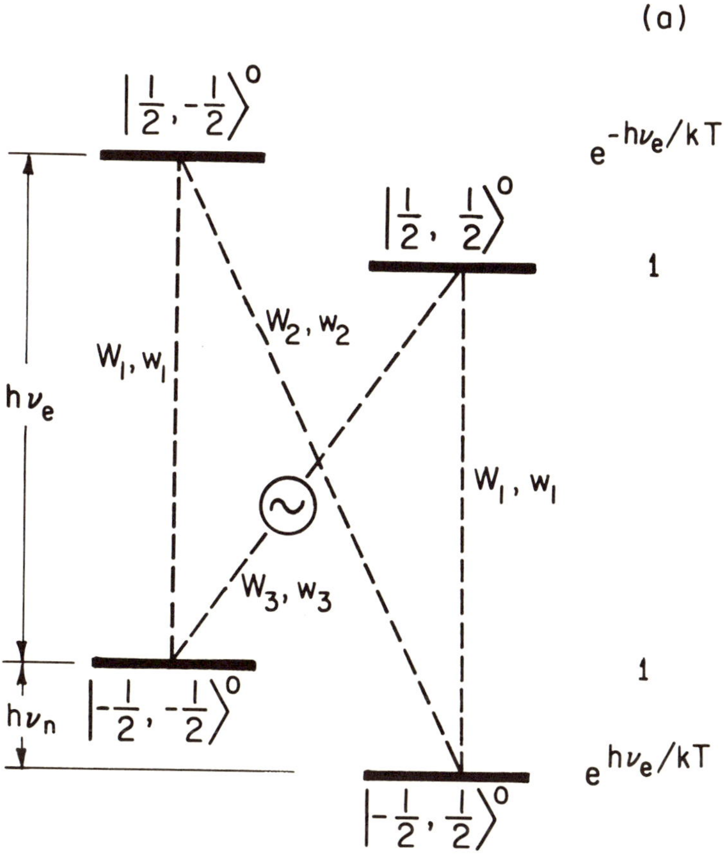

FIG. 5. ENERGY LEVELS AND STATES $|M,m\rangle^\circ$ for nucleus **I** and paramagnetic spin **S** in dipolar coupling in a solid and in a magnetic field. If the forbidden transition W_3 is strongly saturated, the dynamic populations of column (a) result, yielding an enhanced nuclear polarization, Eq. (9). This is a method used in polarizing targets.

lator to $\nu = \nu_e - \nu_n$. Then the relative populations will become those shown in column (a), yielding a large dynamic nuclear polarization

$$p = \tanh(h\nu_e/2kT) \approx h\nu_e/2kT \quad . \tag{9}$$

If W_2 is saturated at $\nu = \nu_e + \nu_n$, the same enhancement is observed but the polarization is reversed. This is crudely what happens between a Nd ion and the near protons at $r \lesssim 10$Å . More distant protons are polarized by this inner shell through nuclear spin diffusion, i.e., mutual nuclear spin flips. Thus eventually all the protons in the crystal become polarized to the value in Eq. (9). After flipping a proton down, the Nd ion is quickly reversed by its strong lattice relaxation w_1; it is then ready to flip down another proton by the applied rf field, etc., and so the few Nd ions are used again and again to polarize the many protons, in a King Solomon fashion.

Some experimental results are shown (figure 6) for Nd:LMN at 4.2K;[17] large positive and negative polarizations are clearly seen at the saturation of transitions W_3 and W_2, respectively. No polarization is observed when the transition W_1 is saturated. At 1.5K a proton polarization of 72% was observed compared to the theoretical value of 83% from Eq. (9). This experiment showed for the first time that large polarizations could be obtained with subsequent application to polarized targets.

A detailed calculation[18] shows that the proton relaxation rate is approximately

$$\frac{1}{T_{1n}} = \frac{\bar{\sigma}\,\mathrm{sech}^2(h\nu_e/2kT)}{T_{1e}} , \tag{10}$$

where $\bar{\sigma}$ is the geometric mean of Eq. (8) over the shell of influence around a Nd ion and T_{1e} is the Nd spin-lattice relaxation time. At low temperatures we have shown theoretically and experimentally[19] that T_{1e} is due to the

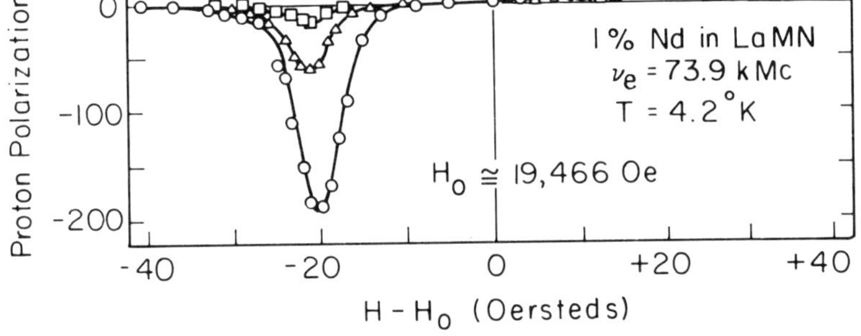

FIG. 6. ENHANCEMENT OF PROTON POLARIZATION in Nd:LMN from reference 17, showing positive and negative proton polarizations as the forbidden transition W_3 or W_2 of figure 5 are selectively saturated at three microwave attenuator levels (db). At lower temperatures (1.5K), polarizations of 70% can be obtained and used in polarized targets.

direct process and becomes temperature independent, limited only by spontaneous phonon emission. The important consequence is that the nuclear relaxation time $T_{1n} \propto \exp[g\beta H/2kT]$ becomes exponentially longer at reduced temperatures—easily many days! This was first observed by Gunter and Jeffries,[20,21] who suggested the possibility of producing "frozen targets" by first dynamically polarizing at a higher temperature and in a required uniform resonance magnetic field, and then holding this polarization at a lower temperature and in a much less uniform and hence more geometrically accessible magnetic field.

The first large polarized target designed for use in the 100 Mev to Gev range was constructed at Berkeley by Chamberlain, Jeffries, Schulz, Shapiro, and Van Rossum.[22] It used four Nd:LMN crystals of volume ≈ 20 cm^3 in the apparatus shown in figure 7. The initial use was for pion-proton scattering at 250 Mev.

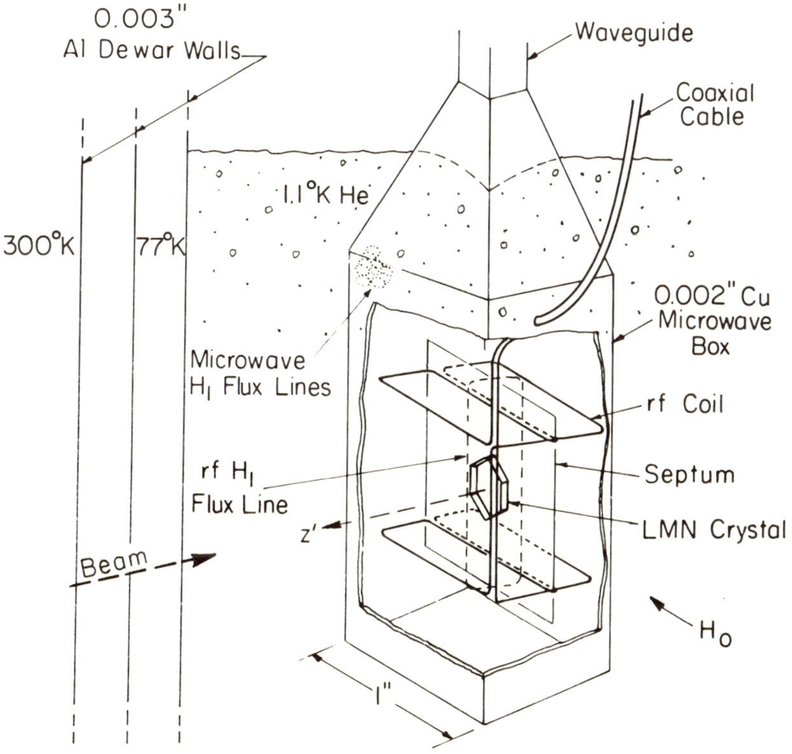

FIG. 7. DETAILS OF THE FIRST LARGE POLARIZED TARGET (Berkeley), reference 22. Only one of four large Nd:LMN crystals is shown.

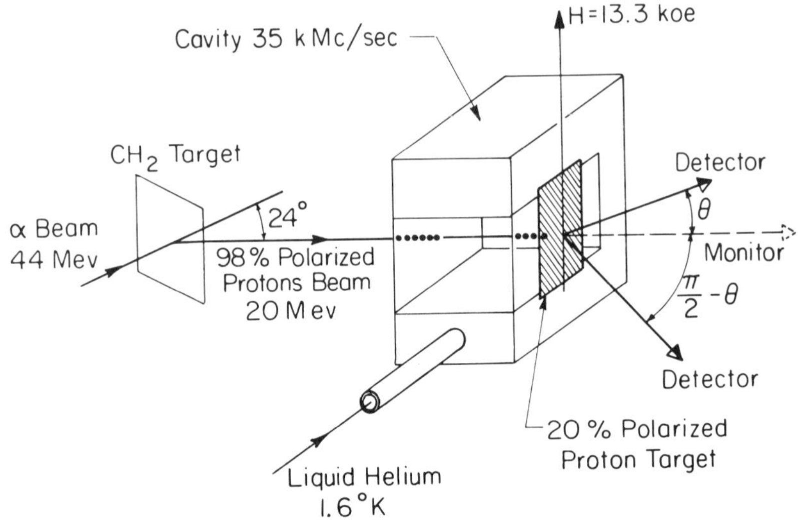

FIG. 8. DETAILS OF THE SACLAY 1962 EXPERIMENT (reference 23), the first experiment using a very thin polarized target of Ce:LMN.

Concurrently with this work, Abragam, Borghini, et al.[23] constructed at Saclay the first very thin polarized proton target and even used it with a polarized *proton beam* at 20 Mev. The Saclay group refers to this method as the "effet solide"; their target (see figure 8) used a Ce:LMN crystal and was used to measure the spin correlation parameter C_{nn}.

By around 1966 Nd:LMN seemed to be the best target substance for protons, and several large targets were constructed,[24] yielding easily reversible polarizations of ~70 to 80% at $T \sim 1$K, $H \approx 20\ kOe$ and $\nu \approx 75$ GHz. But LMN is only 3% by weight hydrogen and has a ratio ϱ = (protons not in H/protons in H) ≈ 15. Obviously protons not in H but in heavier nuclei are not polarized but still scatter. Target materials with higher hydrogen density and lower ϱ value would be much more desirable from the viewpoint of scattering statistics. Since this method of proton polarization requires only a solid hydrogenous material in which is imbedded a small concentration of a paramagnetic species, a large number of other target materials has been investigated:[25] liquid NH_3, or glycerol, or glycol doped with paramagnetic Cr-V complexes; butanol doped with the free radical porphyrexide; 1,2 propanediol doped with Cr-V complexes. These materials have values of ϱ ranging from ~3 to 5. They are doped with the paramagnetic impurity in the liquid state, and then frozen in an elaborate and empirically determined way to achieve a working target that is stable and yields the largest proton polarization. Operating temperatures have been extended to ~0.4K by pumping on He^3, fields to H $\approx$ 50 *kOe*, and frequencies to $\nu \approx 110$ GHz—yielding proton polarizations as high as 98%. Niinikoski[26] has constructed targets using dilution refrigerators; he predicts that future cryogenic technology should produce highly polarized targets of volume $\approx$ 100 liters, a large target indeed! Several frozen targets have been made with a polarization loss of only a few percent per day; deuteron targets with 40% polarization have been achieved.[26]

In the present target materials, organic liquids containing Cr-V complexes or other magnetic species and frozen into a glassy solid, the electron magnetic resonance line is considerably broadened by anisotropic effects, and the transitions W_1, W_2, and W_3 (figure 5) are not well resolved: the forbidden transition cannot really be selectively saturated. Nevertheless at sufficiently large values of H/T, large dynamic polarizations are observed: the mechanism has been most fully explained by Borghini.[27] The theory is complex and there are several special cases. Basically in a broad electron resonance line, rf saturation can induce a cooling of the electron spin-spin sub-reservoir of width $\Delta\nu_e$. This reservoir comes into thermal contact with the nuclear spin reservoir when $\Delta\nu_e \approx \nu_n$, and the nuclei are cooled. It is probable that this mechanism and its variants operate simultaneously with that of figure 5.

I would like to note that there is another method of polarization, the "nuclear spin refrigerator,"[28,29] which does not require any rf pumping: nuclei are polarized in a quantum mechanical type of Carnot cycle by rapid rotation of an anisotropic crystal in a magnetic field. Proton polarizations of 65% have recently been achieved in $(Yb,Y)(C_2H_5SO_4)_3 \cdot 9H_2O$;[30] the method does not require a uniform magnetic field, making the target more geometrically accessible to scattering beams.

The successful technological development of polarized targets is perhaps an unanticipated outgrowth of the discipline of magnetic resonance; Bloch's ideas and contributions can be clearly traced. His early detailed theoretical papers using density matrix methods[31,13,32] influenced the later works of Redfield,[33] Provotorov,[34] and Borghini,[27] which provide the most comprehensive theoretical treatment of phenomena like dynamic polarization. More generally, the entire community of physics is indebted to Felix Bloch for his deep insights and major contributions to so many fields of physics in the past half-century.

ACKNOWLEDGMENT

I wish to acknowledge the support of the Division of Materials Sciences, Office of Basic Energy Sciences, U.S. Department of Energy.

REFERENCES

1. F. Bloch, W. W. Hansen, and M. Packard, Phys. Rev. **70**, 474 (1946).
2. E. M. Purcell, H. C. Torrey, and R. V. Pound, Phys. Rev. **69**, 37 (1946).
3. F. Bloch and C. D. Jeffries, Phys. Rev. **80**, 305 (1950); C. D. Jeffries, Phys. Rev. **81**, 1040 (1951).
4. A. Overhauser, Phys. Rev. **92**, 411 (1953).
5. T. R. Carver and C. P. Slichter, Phys. Rev. **92**, 212 (1953).
6. Carson Jeffries, Phys. Today, April 1954, pp. 30-32.
7. J. Korringa, Phys. Rev. **94**, 1388 (1954).
8. C. Kittel, Phys. Rev. **95**, 589 (1954).
9. P. Brovetto and G. Cini, Nuovo Cimento **11**, 618 (1954); P. Brovetto and S. Ferroni, Nuovo Cimento **12**, 90 (1954).
10. M. J. Klein, Phys. Rev. **98**, 1736 (1955).
11. A. Overhauser, New York Am. Phys. Soc. Meeting, Phys. Rev. **94**, 768 (1954).
12. A. Abragam, Phys. Rev. **98**, 1729 (1955). It is of historical interest for this Festschrift that this paper was written during Abragam's stay at CERN in the same year that Bloch was Director-General, with whom Abragam acknowledges many illuminating discussions.
13. F. Bloch, Phys. Rev. **102**, 104 (1956).
14. C. D. Jeffries, Bull. Am. Phys. Soc. **2**, 65 (1957); C. D. Jeffries, Phys. Rev. **106**, 164 (1957); **117**, 1056 (1960).
15. M. Abraham, R. W. Kedzie, and C. D. Jeffries, Phys. Rev. **106**, 165 (1959).
16. A. Abragam and W. G. Proctor, Compt. rend. **246**, 2253 (1958).
17. T. J. Schmugge and C. D. Jeffries, Phys. Rev. Lett. **9**, 268 (1962); Phys. Rev. **138A**, 1785 (1965); C. D. Jeffries, *Dynamic Nuclear Orientation* (Wiley-Interscience, New York, 1963).
18. C. D. Jeffries, *Electron Paramagnetic Resonance*, ed. S. Geschwind (Plenum Press, New York, 1972), chapter 3.
19. See, e.g., P. L. Scott and C. D. Jeffries, Phys. Rev. **127**, 32 (1962).
20. T. E. Gunter and C. D. Jeffries, Phys. Rev. **159**, 290 (1967).
21. C. D. Jeffries, *Proceedings of the International Conference on Polarized Targets and Ion Sources*, Saclay, France, December 1966 (La Documentation Française, Paris, 1966), p. 95.
22. O. Chamberlain, C. D. Jeffries, C. H. Schultz, G. Shapiro, and L. van Rossum, Phys. Lett. **7**, 293 (1963).
23. A. Abragam, M. Borghini, P. Catillon, J. Coustham, P. Roubeau, and J. Thirion, Phys. Lett. **2**, 310 (1962).
24. See, e.g., *Proceedings of the International Conference on Polarized Targets and Ion Sources*, Saclay, France, December 1966 (La Documentation Française, Paris, 1966); and *Polarization Phenomena in Nuclear Reactions*, ed. H. H. Barschall and W. Haeberli (University of Wisconsin Press, Madison, 1971).
25. *Proceedings of the Second International Conference on Polarized Targets, Berkeley 1971*, ed. G. Shapiro (National Technical Information Service, Springfield, Virginia, 1972), LBL-500, UC-34 Physics.
26. *Syposium on High Energy Physics With Polarized Beams and Targets, Argonne, 1976*, ed. M. L. Marshall (American Institute of Physics, New York, 1976).
27. M. Borghini, Phys. Lett. **26A**, 242 (1968); see also reference 25, p. 1.
28. C. D. Jeffries, Cryogenics **3**, 41 (1963); K. H. Langley and C. D. Jeffries, Phys. Rev. **152**, 358 (1966).
29. A. Abragam, Cryogenics **3**, 42 (1963).

30. J. Button-Shafer, R. L. Lichti, and W. H. Potter, Phys. Rev. Lett. **39**, 677 (1977).
31. R. K. Wangsness and F. Bloch, Phys. Rev. **89**, 728 (1953).
32. F. Bloch, Phys. Rev. **105**, 1206 (1957).
33. A. G. Redfield, IBM J. Res. Dev. **1**, 19 (1957).
34. B. N. Provotorov, Sov. Phys.—JETP **14**, 1126 (1961).

LOCAL THEORY OF ELECTRONIC STRUCTURE OF EXTENDED SYSTEMS

by Walter Kohn and Avishay Yaniv

ABSTRACT

We present a theory of local electronic structure that yields the local density of states $n(\mathbf{r},E)$ and related quantities. The theory applies whether the electronic wave functions themselves are delocalized or not and requires a knowledge of the Hamiltonian only in the vicinity of $\mathbf{r}$. Several simple examples are worked out with good results. Possible applications include periodic and non-periodic transition metal systems and large molecules.

I. INTRODUCTION

One of the most general characteristics of condensed matter, in its normal state, is that the electronic properties in the vicinity of a point $\mathbf{r}$ depend significantly on the electrostatic potential $\phi(\mathbf{r}')$ at points $\mathbf{r}'$ near $\mathbf{r}$. This physical property of "locality" is not directly evident from the wave-mechanical equations that govern such a system. Thus, for example, in the framework of the density functional theory, one is required to solve the wave equations

$$(-\tfrac{1}{2}\nabla^2 + V(\mathbf{r}))\,\psi^\nu(\mathbf{r}) = E^\nu\psi^\nu(\mathbf{r}) \quad , \tag{1.1}$$

where $V(\mathbf{r})$ is an appropriately calculated effective potential. A typical physical property, such as the local single-particle density of states, averaged over a finite energy interval ΔE, is given by a sum over states

---

Walter Kohn is Director, Institute for Theoretical Physics, Santa Barbara, California, and Professor of Physics at the University of California, San Diego. Avishay Yaniv is Assistant Professor of Physics at Tel Aviv University, Israel.

$$\bar{n}(\mathbf{r},E) = \frac{1}{\Delta E} \sum_{E<E^\nu<E+\Delta E} |\psi^\nu(\mathbf{r})|^2 \quad . \tag{1.2}$$

The individual normalized eigenfunctions $\psi^\nu(\mathbf{r})$ are in general not localized, and do not have the locality property mentioned above; i.e., their value at a point $\mathbf{r}$ is, in general, strongly sensitive to changes in $V(\mathbf{r}')$ even for $\mathbf{r}'$ arbitrarily far from $\mathbf{r}$. However, physical quantities involving sums, such as in Eq. (1.2), have been found in practice to have the locality property. As a simple example we mention that calculated physical properties of a solid surface are substantially the same whether one calculates them for a semi-infinite solid or a slab a few layers thick.

In recent years several calculational techniques have been developed and applied, which depend on this locality property. Among these we mention cluster[1,2,3,4] calculations, in which an extensive system is replaced by a small cluster of atoms surrounding the point of interest; and moments (and related) methods[5,6] used in conjunction with the tight binding approximation for the eigenfunctions $\psi^\nu(\mathbf{r})$.

In a recent paper[7] we have proved a strict locality theorem for the Laplace transform of the local density of states (LDS)

$$\varrho(\mathbf{r},\beta) \equiv \int n(\mathbf{r},E)\, e^{-\beta E} dE \quad , \tag{1.3}$$

which we call the local partition function (LPF). We proved that if the potential $V(\mathbf{r}')$ was changed only at distant points $\mathbf{r}'$, with $|x_i' - x_i| > a$, the effect on $\varrho(\mathbf{r},\beta)$ decayed in a Gaussian fashion as a function of a. A similar result was also proved for the more general density matrix $\varrho(\mathbf{r},\mathbf{r}';\beta)$, which reduces to $\varrho(\mathbf{r},\beta)$ for $\mathbf{r}=\mathbf{r}'$.

This suggests that $\varrho(\mathbf{r},\mathbf{r}';\beta)$ and $\varrho(\mathbf{r},\beta)$ could be approximately calculated in a basis of functions, $u_n(\mathbf{r}'')$, localized near the points $\mathbf{r}$ and $\mathbf{r}'$ under consideration, whether the eigenfunctions, $\psi^\nu(\mathbf{r})$, are localized or extended. This will indeed be shown to be possible. The mechanics of the method is similar to the Rayleigh-Ritz method for a finite system. However, except in special cases, it does not yield approximate eigenfunctions. It does yield an arbitrarily close approximation to local physical quantities, such as $\varrho(\mathbf{r},\mathbf{r}';\beta)$ or $\bar{n}(\mathbf{r},E)$.

We shall also point out how this direct method of locally calculating $\bar{n}(\mathbf{r},E)$ (without having to calculate the extended eigenfunctions of the entire extended system) can be used in conjunction with the density functional theory.[8,9] In this way, many-body effects can be incorporated, and quantities such as the density $n(\mathbf{r})$ and the energy density $e(\mathbf{r})$ of extensive systems can be calculated self-consistently. Since, in view of general density functional theorems, all physical properties are functionals of $n(\mathbf{r})$, other quantities—like the many-body density of states, etc.—can in principle be subsequently obtained.

Finally, we report several illustrative examples, free electrons, electrons at interfaces, and hybridized electrons, which demonstrate how the method works in practice.

II. THE BLOCH EQUATION OF MOTION IN A LOCALIZED REPRESENTATION

Clearly, the density matrix $\varrho(\mathbf{r}_1,\mathbf{r}_2;\beta)$ satisfies the so-called Bloch equation of motion

$$\frac{\partial}{\partial\beta}\varrho(\mathbf{r}_1,\mathbf{r}_2;\beta) = -H(\mathbf{r}_2)\,\varrho(\mathbf{r}_1,\mathbf{r}_2;\beta) \tag{2.1}$$

and the initial condition

$$\varrho(\mathbf{r}_1,\mathbf{r}_2;0) = \delta(\mathbf{r}_1-\mathbf{r}_2) \quad . \tag{2.2}$$

For a first orientation, we consider the case of free electrons $H(\mathbf{r}_2) = -\tfrac{1}{2}\nabla_2^2$. Eq. (2.1) becomes

$$\frac{\partial}{\partial\beta}\varrho_o(\mathbf{r}_1,\mathbf{r}_2;\beta) = \tfrac{1}{2}\nabla_2^2\,\varrho_o(\mathbf{r}_1,\mathbf{r}_2;\beta) \quad . \tag{2.3}$$

This is the equation for heat-conduction if β is interpreted as time and ϱ_o as the temperature at $\mathbf{r}_2$, $\mathbf{r}_1$ being an idle parameter. In this case (2.1) and (2.2) represent the heat flow from an initial point source. The solution is given by a spreading Gaussian

$$\varrho_o(\mathbf{r}_1,\mathbf{r}_2;\beta) = \frac{1}{(2\pi\beta)^{3/2}}\exp(-|\mathbf{r}_2-\mathbf{r}_1|^2/2\beta) \quad . \tag{2.4}$$

When the Hamiltonian contains also a potential energy, $V(\mathbf{r})$, the behavior of ϱ for short times, β, is still given by (2.4), since the effect of the kinetic energy operator $\tfrac{1}{2}\,p^2$ is much greater than $V(\mathbf{r})$ in view of the very short wavelength of the Fourier components of ϱ. At longer times, the Bloch equation (2.1) may be regarded as a modified heat diffusion equation. The solution is a wave-packet (in $\mathbf{r}_2$) spreading out from the original point source at $\mathbf{r}_1$.

Suppose we are interested in $\varrho(\mathbf{r}_1,\mathbf{r}_2;\beta)$ for $\mathbf{r}_1$ and $\mathbf{r}_2$ near the origin. Then, in view of the character of ϱ described above, we may expand it in a complete set of functions, u_n, localized near the origin,

$$\varrho(\mathbf{r}_1,\mathbf{r}_2;\beta) = (\mathbf{r}_1|e^{-\beta H}|\mathbf{r}_2) =$$

$$\lim_{N\to\infty} \sum_{n_1,n_2=1}^{N} (\mathbf{r}_1|n_1)(n_1|e^{-\beta H}|n_2)(n_2|\mathbf{r}_2). \tag{2.5}$$

In practice, the sums will have to be terminated at a finite value of N. Let k_N be a typical wave-number in the Fourier transform of u_N. Then, since the characteristic wave-number of the *short* time solution $\varrho_o(\mathbf{r}_1,\mathbf{r}_2;\beta)$ is $\beta^{-1/2}$, an adequate expansion for a short time β requires N to be large enough so that

$$k_N \geq \beta^{-1/2} \qquad (\text{for small } \beta) \quad . \tag{2.6}$$

For *long* times β, we see from the free electron example, Eq. (2.4), that the solution spreads out. Let us denote its spatial dimension by $R(\beta)$ (for free electrons, $R_o(\beta) \sim \sqrt{2\beta}$). Then, if the range of u_N is denoted by R_N, an adequate expansion of ϱ requires

$$R_N \geq R(\beta) \qquad (\text{for large } \beta) \quad . \tag{2.7}$$

Thus, we see that, for a fixed maximum N and corresponding values of k_N and R_N, the representation (2.5) will fail for both very small and very large β, when the conditions (2.6) and (2.7) cease to be satisfied. However, for intermediate values of β, the expansion (2.5), up to a finite large N, will be satisfactory.

Here we also note that the choice of the length scale a of the functions u_n is at our disposal. Clearly

$$k_N \propto a^{-1}, \; R_N \propto a \quad . \tag{2.8}$$

Thus, for a given N, one can choose a so as to obtain a better representation of ϱ for either small β (choose a small) or large β (choose a large).

The representation (2.5) of ϱ is inconvenient because of the appearance of the operator H in the exponential. To obtain a practical form for ϱ, we write

$$H = H_N + \Delta H_N \quad , \tag{2.9}$$

where H_N is the projection of H on the N functions u_n,

$$H_N \equiv \sum_{n_1,n_2=1}^{N} |n_1)(n_1|H|n_2)(n_2| \quad . \tag{2.10}$$

Clearly, if N functions $\{u_n\}$ are adequate for a description of ϱ, the projected operator, H_n, will be an adequate representation of H. Thus, we may replace Eq. (2.5) by:

$$\varrho_N(\mathbf{r}_1,\mathbf{r}_2;\beta) = \sum_{n_1,n_2=1}^{N} (\mathbf{r}_1|n_1)(n_1|e^{-\beta H_N}|n_2)(n_2|\mathbf{r}_2) \quad . \qquad (2.11)$$

This can be transformed into a more convenient expression by diagonalizing H_N,

$$\sum_{n_2=1}^{N} (n_1|H|n_2)\, c^{\nu}_{n_2} = E^{\nu} c^{\nu}_{n_1} \qquad (2.12)$$

with eigenvectors

$$\phi^{\nu}(\mathbf{r}) = \sum_{n=1}^{N} c^{\nu}_n u_n(\mathbf{r}) \quad . \qquad (2.13)$$

Then, because of the fact that the $\{\phi^{\nu}\}$ arise from the $\{u_n\}$ by a unitary transformation, we can rewrite Eq. (2.11) as:

$$\varrho_N(\mathbf{r}_1,\mathbf{r}_2;\beta) = \sum_{\mu_1\mu_2=1}^{N} (\mathbf{r}_1|\mu_1)(\mu_1|e^{-\beta H_N}|\mu_2)(\mu_2|\mathbf{r}_2)$$

$$= \sum_{\mu=1}^{N} \phi^{\mu}(\mathbf{r}_1)^* \phi^{\mu}(\mathbf{r}_2) e^{-\beta E^{\mu}} \quad . \qquad (2.14)$$

This is the approximate representation of ϱ in terms of the u_n's.

III. THE DENSITY MATRIX AND THE DENSITY OF STATES AS A FUNCTION OF E

From Eq. (2.14) we can immediately obtain, by Laplace inversion, an approximation to the density matrix for a given energy E,

$$n_N(\mathbf{r}_1,\mathbf{r}_2;E) = \sum_{\mu=1}^{N} \phi^{\mu}(\mathbf{r}_1)^* \phi^{\mu}(\mathbf{r}_2)\, \delta(E-E^{\mu}) \quad , \qquad (3.1)$$

and, in particular, the local density of states

$$n_N(\mathbf{r},E) = n_N(\mathbf{r},\mathbf{r};E) = \sum_{\mu=1}^{N} |\phi^\mu(\mathbf{r})|^2 \delta(E-E^\mu) \quad . \tag{3.2}$$

We note that, in this procedure, $n_N(\mathbf{r},E)$ is a sum of δ-functions in E, even when the real density of states $n(\mathbf{r},E)$ is continuous. A well known way of smoothing (3.2) is to calculate the integrated function

$$\mathcal{N}_N(\mathbf{r},E) = \int^E n_N(\mathbf{r},E')dE'$$

$$= \sum_{E^\mu \leq E} |\phi^\mu(\mathbf{r})|^2 \quad , \tag{3.3}$$

which is a stepwise monotonically increasing function of E, smooth this function to yield $\overline{\mathcal{N}}_N(\mathbf{r},E)$, and then obtain the smoothed $\overline{n}_N(\mathbf{r},E)$ by differentiation with respect to E.

IV. FREE ELECTRONS IN A LOCALIZED REPRESENTATION

In this section we shall discuss the approximate representation of the density matrix $n(x,x';E)$, of a one-dimensional free electron gas in a basis of a finite number of harmonic oscillator functions.

The free electron Hamiltonian is

$$H_o = \tfrac{1}{2} p^2 \quad . \tag{4.1}$$

The local density of states is, in this case, independent of the position x,

$$n(x,E) = n(E) = \frac{1}{\pi}(2E)^{-\frac{1}{2}} \quad , \tag{4.2}$$

and the corresponding integrated density of states is

$$\mathcal{N}(x,E) = \mathcal{N}(E) = \frac{1}{\pi}(2E)^{\frac{1}{2}} \quad . \tag{4.3}$$

To calculate $n(x,E)$ at the origin, we use as basis functions the well known eigenfunctions u_n of the harmonic oscillator Hamiltonian,

$$H = \tfrac{1}{2} p^2 + \tfrac{1}{2}\omega^2 x^2 \quad , \tag{4.4}$$

with eigenvalues

$$E_n = \omega(n + ½) \quad . \tag{4.5}$$

In this representation the matrix elements of H_o are

$$(m|H_o|n) = \frac{\omega}{4}\Big[-\sqrt{(n-1)n}\,\delta_{m,n-2} + (2n+1)\,\delta_{m,n} - \sqrt{(n+1)(n+2)}\,\delta_{m,n+2}\Big] \quad . \tag{4.6}$$

Clearly, these matrix elements vanish unless m and n have the same parity. To calculate $n(E) = n(o,E)$, we can restrict ourselves to the subspace of even functions only. We have taken $\omega = 1$, and numerically diagonalized the matrix (4.6) for two cases. In the first case we used 5 functions and in the second 10. From this we have calculated $n_N(E)$ and $\mathcal{N}_N(E)$, using Eqs. (3.2) and (3.3). Figures 1a and 1b compare the results with the exact $\mathcal{N}(E)$, Eq. (4.3). Also shown in these figures are the midpoints of each vertical rise in the approximate integrated density of states. As can be seen, the agreement is very good up to $E \approx 6$ for $N = 5$ and up to $E \approx 15$ for $N = 10$. (In this case the energies scale as ω.) In the Appendix, we prove that as N becomes large, the eigenvalues E^ν become dense, and the approximate integrated density of states converges to the exact result, Eq. (4.3).

V. SURFACES AND INTERFACES IN A LOCALIZED REPRESENTATION

As a second example of the application of our procedure, we derive the density distribution of the so-called sine-gas, near a surface created by an infinite potential barrier at $x = 0$. We assume the electron gas to occupy the half-space $x \geq 0$. To solve this problem within our formalism, one needs a localized basis near $\mathbf{r} = 0$, which also vanishes at $x = 0$. Two possible bases that fulfill these requirements are for example

$$\psi_{lmn}(\mathbf{r}) = \sqrt{2}\, u_l(z)\, u_m(y)\, u_{2n-1}(x) \tag{5.1}$$

and

$$\tilde{\psi}_{lmn}(\mathbf{r}) = \left(\frac{2l!\,\omega^{m+1}}{(l+m)!\,\pi}\right)^{½} e^{im\phi}\,\varrho^m\, e^{-½\omega\varrho^2}\, L_l^m(\omega\varrho^2)\, u_{2n-1}(x) \quad , \tag{5.2}$$

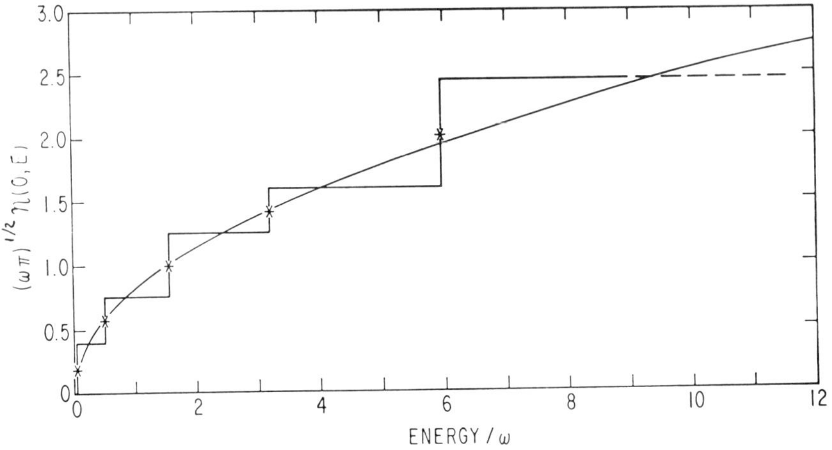

a

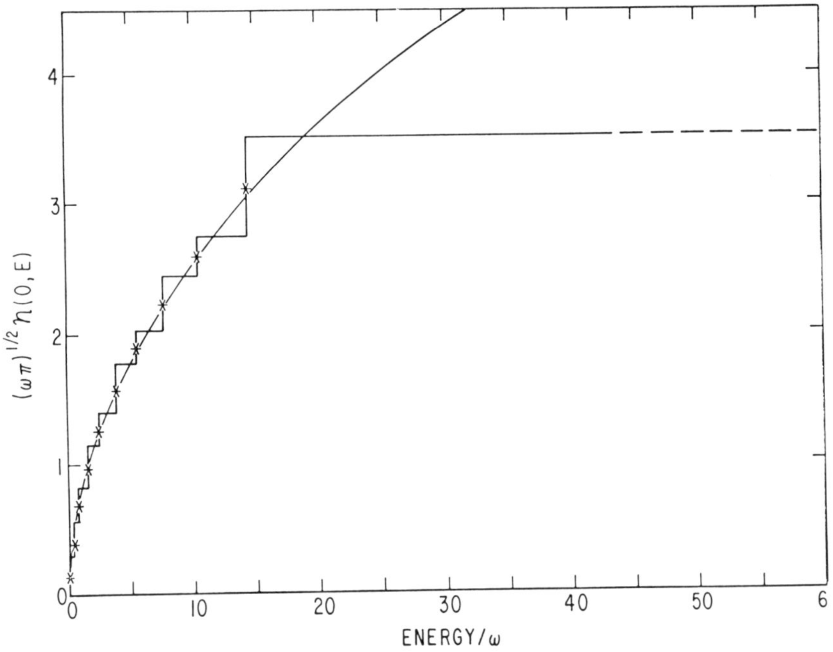

b

FIG. 1. A COMPARISON OF THE APPROXIMATE FREE-ELECTRON DENSITY OF STATES (the staircase curve) with the exact result (the smooth curve). (1a) using 5 even oscillator functions, (1b) using 10 even oscillator functions.

where u_n and L_l^m are the harmonic oscillator wave-functions and the associated Laguerre polynomials respectively. The first basis, (5.1), is convenient for problems having cubic symmetry, whereas (5.2) is more suitable for problems with cylindrical symmetry. For the sine-gas, both (5.1) and (5.2) can be used. Since the problem under consideration has translational symmetry parallel to the surface, it is enough to calculate the density distribution for $y=z=0$. Thus, we can further simplify the problem by using only even l and m when using the basis (5.1), or set $m=0$ when using (5.2). Figures 2a and 2b show the density profile near the surface, calculated in both bases, (5.1) and (5.2). Figure 2a shows the result of a calculation done with a set of $10 \times 10 \times 10$ functions in the x, y, and z directions, respectively. Figure 2b shows a similar result obtained by using 20 cylindrical functions and 10 functions in the x direction (using the basis (5.2)). As can be seen from these figures, the accuracy of the calculation is quite good near the surface, where the basis functions are localized, and becomes less and less satisfactory as one goes further from the surface.

Our next example is a simple model of a bimetallic interface. This is a free electron gas, moving in a potential field having the form of a step function. The position of the step is the position of the interface, whereas

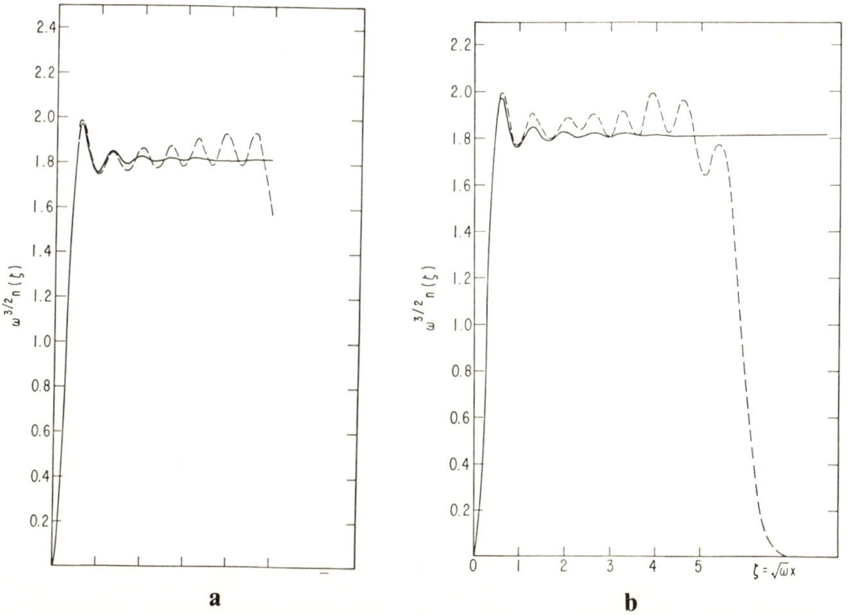

FIG. 2. THE DENSITY PROFILE NEAR THE SURFACE OF A SINE-GAS (solid line), having a Fermi energy of 11.3 ω. The approximate density profile is shown by a dashed line. (2a) is the result obtained by using $10 \times 10 \times 10$ functions—see Eq. (5.1). (2b) was obtained by using 20 cylindrical and 10 perpendicular functions.

the step height is equal to the difference between the Fermi levels of the two metals. Figures 3a and 3b show the density profile near such an interface formed by two metals whose Fermi energies differ by 5 eV and whose bulk electron density ratio is equal to 2. The localized basis used to solve this problem is given by

$$\psi_{l,n}(\mathbf{r}) = \left(\frac{\omega}{\pi}\right)^{1/2} e^{-1/2\omega\varrho^2} L_l(\omega\varrho^2)\, u_n(x) \quad , \tag{5.3}$$

which is localized at the interface. Figure 3a shows the result obtained by using 20 functions in the direction perpendicular to the interface, and 20 cylindrical functions parallel to the interface. Figure 3b shows the results of a similar calculation, obtained by using 40 functions in the perpendicular direction and 20 cylindrical functions. Here again, the accuracy of the calculation is best near the localization center, the interface in this case, and becomes worse away from that point.

VI. *s-d* HYBRIDIZATION IN A LOCALIZED REPRESENTATION

Our last example is the application of a localized basis to the problem of *s-d* hybridization in copper along the [100] direction. In choosing the parameters of our model we follow the work of Mueller[10] on the combined interpolation scheme for copper. Along the [100] direction we have effectively a 1-dimensional problem, with a lattice spacing equal to $a = \frac{1}{2}a_o$ where a_o is the cubic parameter of f.c.c. copper, i.e., $a = 1.81$ Å. Since the five *d*-bands do not mix in this symmetry direction, it is sufficient to consider the hybridization of one *d*-band at a time. We describe the *d*-band before hybridization as a tight-binding band, arising from Wannier functions, ϕ_n, localized at the various lattice sites Z_n, and having the symmetry $(3z^2 - \mathbf{r}^2)$. The Bloch functions and energies of this band are given, respectively, by

$$\psi_d = N^{-1/2} \sum_n e^{ikZ_n} \phi_n \tag{6.1}$$

$$E_d(k) = E_o + 2T\cos(ka) \quad , \tag{6.2}$$

where N is the number of atoms in the 1-dimensional problem, and E_o and T are the self-energy and the transfer integral, respectively. Using the same Koster-Slater parameters as those used by Mueller, we obtain $T = 0.386$ eV.

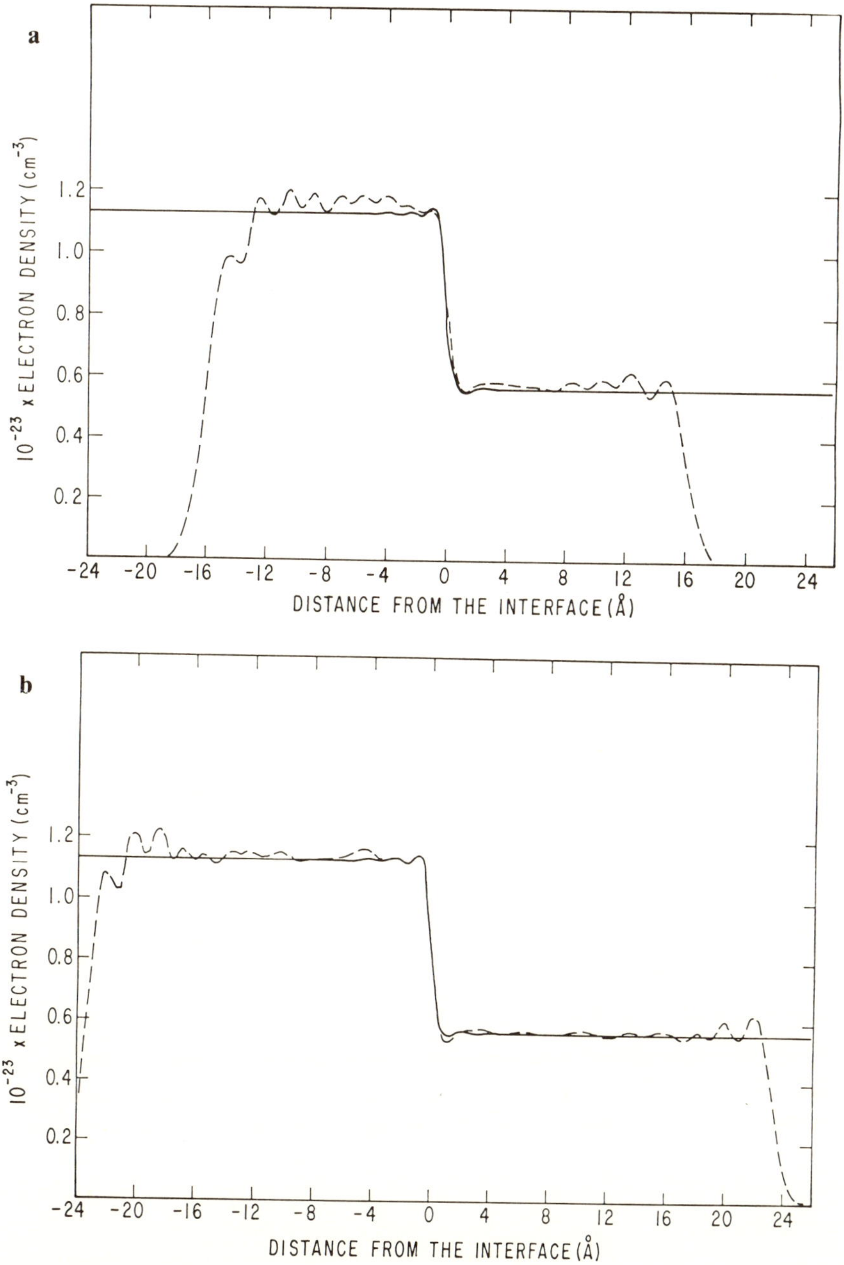

FIG. 3. THE EXACT (solid line) AND THE APPROXIMATE (dashed line) DENSITY DISTRIBUTION near a bimetallic interface. The result of (3a) was obtained by using 20 cylindrical and 20 perpendicular functions, (3b) was obtained by using 20 × 40 functions.

For E_o we take the actual difference between the bottom of the d-band and the conduction band, i.e., $E_o = 4.67$ eV. Thus, the d-electron energy band is given by

$$E_d(k) = 4.67 + 0.772\cos(ka) \qquad \text{(eV)} \quad . \tag{6.3}$$

For the s-electron wave functions and energies, we use plane waves with an effective mass equal to the copper effective mass in the [100] direction,

$$\psi_\nu = (Na)^{-1/2}\, e^{i(k+K_\nu)z} \tag{6.4}$$

$$E_\nu(k) = 1.154\,[a(k+K_\nu)]^2 \qquad \text{(eV)} \quad , \tag{6.5}$$

where K_ν are the vectors of the reciprocal lattice. We assume that the plane waves (6.4) are orthogonal to the Wannier functions ϕ_n.

For the hybridization term, which couples the s and d electrons, we write

$$g_\nu(k) = \gamma\{\exp(-\alpha[a(k+K_\nu)]^2) - \exp(-\beta[a(k+K_\nu)]^2)\} \tag{6.6}$$

where the parameters α, β, and γ are chosen to simulate Mueller's hybridization function. The coefficient α determines the effective range of $g_\nu(k)$ in k-space. By taking $\alpha = 1.2665 \cdot 10^{-2}$ we obtain a range similar to that of Mueller's. To determine β we require $g(k)$ to have its maximum at the Brillouin zone boundary. This gives $\beta = 0.3485$. For γ we take $\gamma = 1.98$ eV so that $g(k)$ is equal to half the hybridization gap at the crossover point of the s and the d bands before hybridization. To summarize, the Hamiltonian of the hybridized bands is given by

$$(\psi_d|H|\psi_d) = E_d(k) \quad , \tag{6.7.a}$$

$$(\psi_\nu|H|\psi_\nu) = E_\nu(k) \quad , \tag{6.7.b}$$

$$(\psi_\nu|H|\psi_d) = (\psi_d|H|\psi_\nu) = g_\nu(k) \quad . \tag{6.7.c}$$

The band structure of this Hamiltonian, obtained by a numerical diagonalization of the matrix (6.7) for various values of k, is shown in figure 4.

Let $\psi^\alpha(k)$ be the Bloch function of the α hybridized band. This wave function can be written as a linear combination of the non-hybridized wave functions

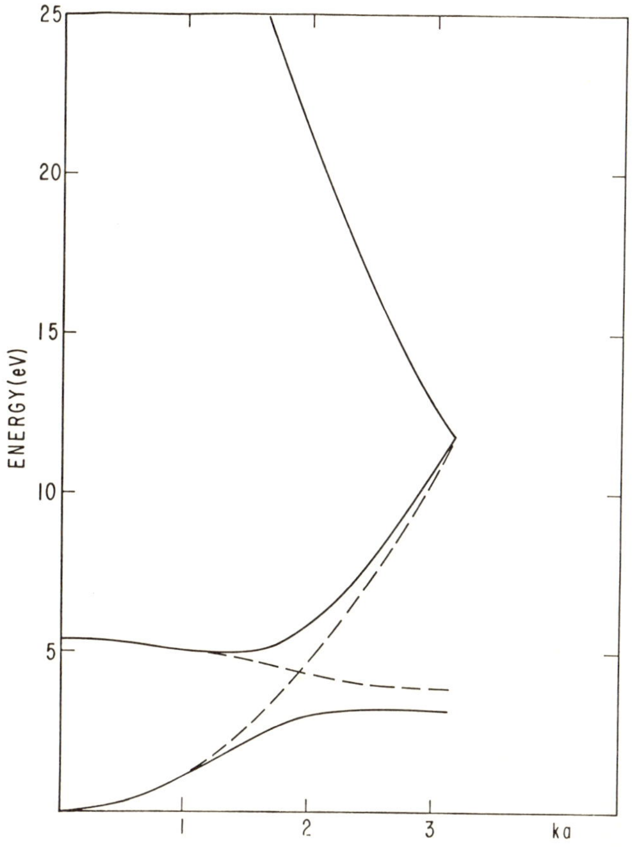

FIG. 4. THE BAND STRUCTURE OF THE HYBRIDIZED COPPER HAMILTONIAN. Shown by a dashed line is the band structure before hybridization.

$$\psi^{\alpha}(k) = c^{\alpha}(k)\psi_d + \sum_{\nu} a_{\nu}^{\alpha}(k)\psi_{\nu} \quad , \tag{6.8}$$

where c^{α} and a_{ν}^{α} are the corresponding expansion coefficients. From the equation above we can define the d density of states per atom as

$$n_d(E) = N^{-1} \sum_{\alpha,k} |c^{\alpha}(k)|^2 \delta(E - E^{\alpha}(k))$$

$$= \frac{a}{\pi} \sum_{\alpha} \left. \frac{|c^{\alpha}(k)|^2}{|dE^{\alpha}(k)/dk|} \right|_{E^{\alpha}(k)=E} \quad , \tag{6.9}$$

where $E^{\alpha}(k)$ is the energy of the α hybridized band. The corresponding integrated density of states is given by

$$\mathcal{N}_d(E) = \int_{-\infty}^{E} n_d(E')dE' = \frac{a}{\pi}\sum_{\alpha}\int_{o}^{\pi/a} |c^{\alpha}(k)|^2\theta(E-E^{\alpha}(k))dk, \quad (6.10)$$

where θ is the unit step function.

In a similar way we can also define the local s-density of states, and the corresponding integrated density, by

$$n_s(z,E) = a\sum_{\alpha,k} \Big|\sum_{\nu} a_{\nu}^{\alpha}(k)\psi_{\nu}(z)\Big|^2\delta(E-E^{\alpha}(k))$$

$$= \frac{a}{\pi}\sum_{\alpha,\mu,\nu}\frac{a_{\nu}^{\alpha}(k)\,a_{\nu}^{\alpha}(k)^*}{|dE^{\alpha}(k)/dk|}e^{i(K_{\nu}-K_{\mu})z}\Bigg|_{E^{\alpha}(k)=E} \quad (6.11)$$

and

$$\mathcal{N}_s(z,E) = \quad (6.12)$$

$$\frac{a}{\pi}\sum_{\alpha,\mu,\nu}\int_{o}^{\pi/a} a_{\nu}^{\alpha}(k)\,a_{\mu}^{\alpha}(k)^* e^{i(K_{\nu}-K_{\mu})z}\theta(E-E^{\alpha}(k))dk \quad .$$

To solve the hybridization model in a localized basis, we use the Wannier functions $\{\phi_n\}$ and the harmonic oscillator functions $\{u_{\nu}\}$. We assume that the set $\{u_{\nu}\}$ is orthogonal to $\{\phi_n\}$ (as was the set of plane waves $\{\psi_{\nu}\}$). In this basis, the hybridized Hamiltonian is given by

$$(\phi_n|H|\phi_n) = E_o \quad ,$$

$$(\phi_n|H|\phi_{n\pm 1}) = T \quad ,$$

$$(u_{\nu}|H|u_{\mu}) = \frac{\hbar^2\omega}{4m^*}\Big[-\sqrt{\mu(\mu-1)}\,\delta_{\nu,\mu-2} + (2\mu+1)\delta_{\mu,\nu}$$

$$-\sqrt{(\mu+1)(\mu+2)}\,\delta_{\nu,\mu+2}\Big] \quad ,$$

$$(u_{2\nu}|H|\phi_n) = a^{1/2}\gamma\left[\frac{1}{(1+2\alpha a^2\omega)^{1/2}}\left(\frac{1-2\alpha a^2\omega}{1+2\alpha a^2\omega}\right)^{\nu}\right.$$

$$\exp\left(\frac{\alpha a^2\omega^2 Z_n^2}{1-4\alpha^2 a^4\omega^2}\right)\times u_{2\nu}\left(Z_n\sqrt{\frac{\omega}{1-4\alpha^2a^4\omega^2}}\right)$$

$$-\frac{1}{(1+2\beta a^2\omega)^{1/2}}\left(\frac{1-2\beta a^2\omega}{1+2\beta a^2\omega}\right)^{\nu}\exp\left(\frac{\beta a^2\omega^2 Z_n^2}{1-4\beta^2a^4\omega^2}\right)\times$$

$$\left.u_{2\nu}\left(Z_n\sqrt{\frac{\omega}{1-4\beta^2a^4\omega^2}}\right)\right],$$

$$(u_{2\nu+1}|H|\phi_n) = a^{1/2}\gamma\left[\frac{1}{(1+2\alpha a^2\omega)^{1/2}}\left(\frac{1-2\alpha a^2\omega}{1+2\alpha a^2\omega}\right)^{\nu+1/2}\right.$$

$$\exp\left(\frac{\alpha a^2\omega^2 Z_n^2}{1-4\alpha^2 a^4\omega^2}\right)\times u_{2\nu+1}\left(Z_n\sqrt{\frac{\omega}{1-4\alpha^2a^4\omega^2}}\right)$$

$$-\frac{1}{(1+2\beta a^2\omega)^{1/2}}\left(\frac{1-2\beta a^2\omega}{1+2\beta a^2\omega}\right)^{\nu+1/2}\exp\left(\frac{\beta a^2\omega^2 Z_n^2}{1-4\beta^2a^4\omega^2}\right)\times \tag{6.13}$$

$$\left.u_{2\nu+1}\left(Z_n\sqrt{\frac{\omega}{1-4\beta^2a^4\omega^2}}\right)\right],$$

where $Z_n = na$, $n = 0, \pm 1, \pm 2, \ldots$ are the lattice sites, and ω is the frequency of the oscillator. For the present calculation we have taken $\omega = a^{-2}$. For practical purposes we have also taken a finite linear chain, with an odd number of atoms, and calculated the density of states at the central atom. Figure 5 shows the integrated *d* density of state, whereas figure 6 shows the corresponding integrated *s* local density of states at the origin. Both calculations were done with a chain of 41 atoms and using 40 harmonic oscillator functions. As can be seen from these figures, the agreement with the exact results is quite good.

VII. DISCUSSION

In this paper we have shown how local electronic structure in an ex-

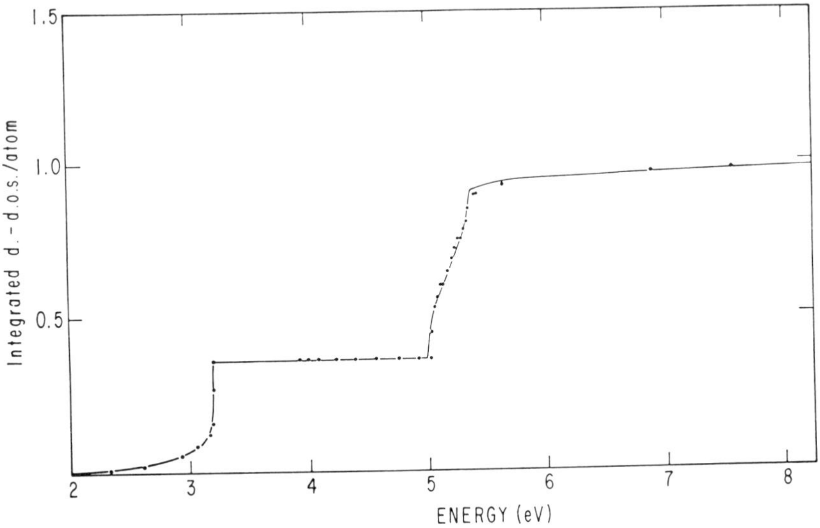

FIG. 5. THE INTEGRATED *d*-DENSITY OF STATES. The points • were obtained using a chain of 41 atoms and 40 oscillator functions.

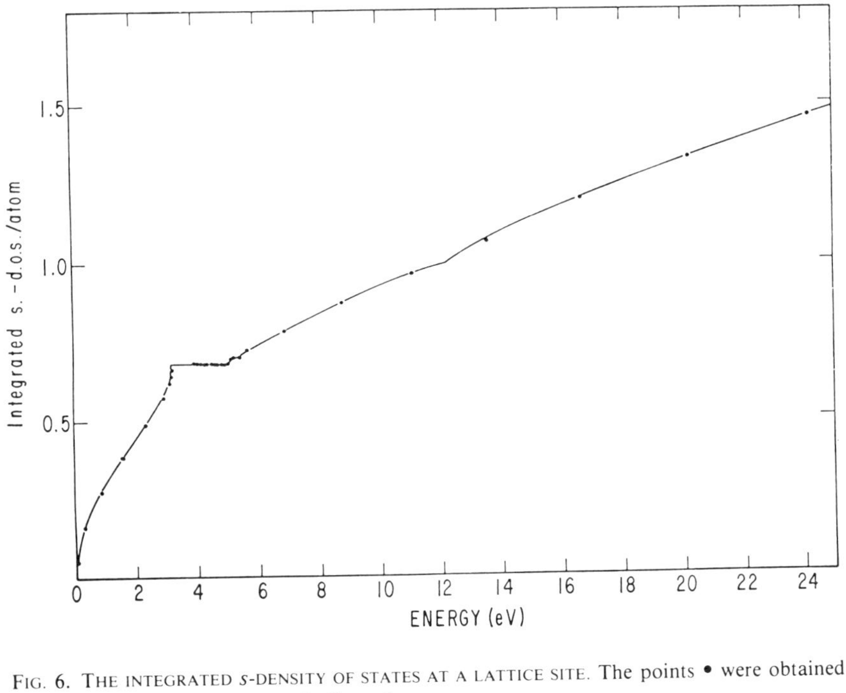

FIG. 6. THE INTEGRATED *s*-DENSITY OF STATES AT A LATTICE SITE. The points • were obtained using the same approximation as in figure 5.

tended system can be calculated in a basis of localized functions, even when the actual eigenfunctions are delocalized. In particular, given any effective potential $V_{eff}(\mathbf{r})$, the total electron density, $n(\mathbf{r})$, can be obtained by integration of $n(\mathbf{r},E)$ over E. In this way, the method can be naturally integrated into the density functional formalism,[8,9] which includes many-body effects.

This method would seem to be especially appropriate for localized defects, such as the surface and interface systems treated in section V, even for hybridized wavefunctions, such as those discussed in section VI. A successful self-consistent calculation for a vacancy, using our method, has been carried out by F. Perrot.[11] Other possible applications are local portions of large molecules.

APPENDIX

In this Appendix we prove the convergence of the procedure of section IV for the one-dimensional free electron gas. To this end we shall calculate explicitly the eigenvectors and the eigenvalues of the truncated Hamiltonian (Eq. (4.6)) in the harmonic oscillator basis. We shall show that as the size of the Hamiltonian matrix is increased, the eigenvalues become dense, and the integrated approximate local density of states, evaluated by using Eq. (3.3), approaches the exact result.

The free electron Hamiltonian has the representation (4.6) in the harmonic oscillator basis. We now approximate this Hamiltonian by a finite matrix of order $2N \times 2N$, whose matrix elements are given by (4.6) with $m,n = 0,1 \ldots 2N-1$. This Hamiltonian can be written as a direct sum of two tridiagonal matrices by separating the even and the odd functions. In this way we have

$$H_o = H_o^- + H_o^+ \quad , \tag{A.1}$$

where H_o^- and H_o^+ are the corresponding representations of H_o in the even and the odd subspaces, respectively. Thus, the problem of diagonalizing the $2N \times 2N$ matrix (4.6) can be separated into two independent problems of diagonalizing the smaller two $N \times N$ matrices $H_o^{\mp}$. Using Eq. (4.6), these matrices are given explicitly by

$$(m|H_o^-|n) \equiv (2m|H_o|2n) =$$

$$\frac{\omega}{4}\left[-\sqrt{(2n-1)2n}\,\delta_{m,n+1} + (4n+1)\delta_{m,n} - \sqrt{(2n+1)(2n+2)}\delta_{m,n+1}\right]$$

$$m,n = 0,1,\ldots N-1 \quad ; \tag{A.2}$$

$$(m|H_o^{+}|n) \equiv (2m-1|H_o|2n-1) =$$
$$\frac{\omega}{4}\left[-\sqrt{(2n-2)(2n-1)}\delta_{m,n+1} + (4n-1)\delta_{m,n} - \sqrt{2n(2n+1)}\delta_{m,n+1}\right]$$
$$m, n = 1, 2, \ldots N \quad . \tag{A.3}$$

It is well known that the characteristic equation of a symmetric tridiagonal matrix can be generated by a recursion relation.[12] In this way, the leading principal minors, of order n, $B_n^{\mp}(y)$ of the matrix $H_o^{\mp} - yI$, obey the recursion relation

$$B_{n+1}^{\mp}(y) = (4n+2\mp 1-y)B_n^{\mp}(y) - 2n(2n\mp 1)B_{n-1}^{\mp}(y)$$
$$B_o^{\mp}(y) = 1$$
$$B_{-1}^{\mp}(y) = 0 \tag{A.4}$$

where y is the dimensionless energy variable,

$$y = 4E/\omega \quad . \tag{A.5}$$

By comparing the recursion relation (A.4) to the recursion relation obeyed by the associated Laguerre polynomials, one can verify that the polynomials $B_n^{\mp}(y)$ are related to the well known polynomials $L_n^{\mp 1/2}(y)$. In fact we have

$$B_n^{\mp}(y) = 2^n n!\, L_n^{\mp 1/2}(y/2) \quad . \tag{A.6}$$

Using the relation between $L_n^{\mp 1/2}$ and the Hermite polynomials we can also write

$$B_n^{-}(y) = \frac{(-1)^n}{2^n} H_{2n}(\sqrt{y/2}) \quad , \tag{A.7}$$

$$B_n^{+}(y) = \frac{(-1)^n}{2^{n+1}\sqrt{y/2}} H_{2n+1}(\sqrt{y/2}) \quad . \tag{A.8}$$

The eigenvalues of the truncated Hamiltonian are given by the roots of $B_N^{\mp}(y)$. As can be seen from Eqs. (A.7) and (A.8), these eigenvalues are simply related to the positive roots of the Hermite polynomials of order $2N$

and $2N+1$. For large N we can use the asymptotic expression[13] for the Hermite polynomials to obtain

$$y_\nu^- = \frac{2\pi^2(\nu - ½)^2}{4N+1} + 0\,(N^{-3/2}) \quad , \tag{A.9}$$

$$y_\nu^+ = \frac{2\pi^2\nu^2}{4N+3} + 0\,(N^{-3/2}) \quad , \tag{A.10}$$

where $y_\nu^\mp$ are the corresponding eigenvalues. As can be seen from these equations, the eigenvalues $y_\nu^\mp$ become dense on the positive y_ν axis as the value of N is increased. In fact, we have

$$y_\nu - y_{\nu-1} \sim \pi\left(\frac{2\,y_\nu}{N}\right)^{½} \quad . \tag{A.11}$$

We turn now to the determination of the eigenvectors of the approximate Hamiltonian. Since the matrices $H_o^\mp$ are tridiagonal, the unnormalized eigenvectors, corresponding to the eigenvalue y_ν, can be written explicitly.[12] Using the expressions derived for $B_n^\mp(y)$ and the expression for the off-diagonal matrix elements of $H_o^\mp$, we obtain

$$c_m^{-\nu} = \frac{(-1)^m}{2^m\sqrt{2m!}}\,H_{2m}(\sqrt{y_\nu^-/2})$$

$$m = 0,1,\ldots \quad N-1 \tag{A.12}$$

and

$$c_m^{+\nu} = \frac{(-1)^{m-1}}{2^m}\left(\frac{2}{(2m-1)!\,y_\nu^+}\right)^{½} H_{2m-1}(\sqrt{y_\nu^+/2})$$

$$m = 1,2,\ldots \quad N \quad . \tag{A.13}$$

In order to normalize the eigenvectors we have to evaluate the following sums:

$$N_\nu^- = \sum_{m=0}^{N-1} (c_m^{-\nu})^2 = \sum_{m=0}^{N-1} \frac{H_{2m-1}^2(\sqrt{y_\nu^-/2})}{2^{2m}(2m)!} \tag{A.14}$$

and

$$N_\nu^+ = \sum_{m=1}^{N} (c_m^{+\nu})^2 = \frac{1}{y_n^+} \sum_{m=1}^{N} \frac{H_{2m-1}^2(\sqrt{y_\nu^+/2})}{2^{2m-1}(2m-1)!} \quad . \tag{A.15}$$

By using the corresponding recursion relations, one can show that any family of orthonormal polynomials, $\{P_n(x)\}$ with respect to a symmetric weight function, obey the following summation identities:

$$\sum_{m=0}^{n} P_{2m}^2(x) = \frac{P_{2n+2}'(x)\,P_{2n}(x) - P_{2n}'(x)\,P_{2n+2}(x)}{2\,A_{2m+2}x} \tag{A.16}$$

$$\sum_{m=1}^{n} P_{2m-1}^2(x) = \frac{P_{2n+1}'(x)\,P_{2n-1}(x) - P_{2n-1}'(x)\,P_{2n+1}(x)}{2\,A_{2m+1}x} \tag{A.17}$$

where

$$A_n = k_n/k_{n-2} \quad , \tag{A.18}$$

and k_n is the coefficient of x^n in $P_n(x)$. Using these identities in the case of the Hermite polynomials gives, after several manipulations,

$$N_\nu^- = \frac{N}{2^{2N-1}(2N-1)!} H_{2N-1}^2(\sqrt{y_\nu^-/2}) \quad , \tag{A.19}$$

$$N_\nu^+ = \frac{(2N+1)}{(y_\nu^+/2)\,2^{2N+2}(2N)!} H_{2N}^2(\sqrt{y_\nu^+/2}) \quad . \tag{A.20}$$

For large N we can use the asymptotic behavior of the Hermite polynomials to obtain

$$N_\nu^- \sim \left(\frac{N}{\pi}\right)^{1/2} \exp(y_\nu^-/2) \tag{A.21}$$

$$N_\nu^+ \sim \left(\frac{N}{\pi}\right)^{1/2} \frac{\exp(y_\nu^+/2)}{y_\nu^+} \tag{A.22}$$

To complete our proof, we evaluate now the residue associated with the various eigenvalues $R_\nu^\mp(x)$, which gives the corresponding jump in the integrated local density of states. Using Eqs. (A.12)-(A.13) and (A.19)-(A.20), we can write

$$R_\nu^-(x) = (N_\nu^-)^{-1}\left[\sum_{m=0}^{N-1} c_m^{-\nu}\, u_{2n}(x)\right]^2$$

$$= \frac{\pi^{1/2} e^{y_\nu^-/2}}{N_\nu^-}\left[\sum_{m=0}^{N-1} (-1)^m u_{2m}(x)\, u_{2m}\left(\sqrt{y_\nu^-/2}\right)\right]^2 , \quad (A.23)$$

$$R_\nu^+(x) = (N_\nu^+)^{-1}\left[\sum_{m=1}^{N} c_m^{+\nu}\, u_{2m-1}(x)\right]^2$$

$$= \frac{\pi^{1/2} e^{y_\nu^+/2}}{N_\nu^+ y_\nu^+}\left[\sum_{m=1}^{N} (-1)^m u_{2m-1}(x)\, u_{2m-1}\left(\sqrt{y_\nu^+/2}\right)\right]^2 , \quad (A.24)$$

where $u_n(x)$ are the normalized harmonic oscillator wavefunctions.

It is not difficult to show that the sums that appear in Eqs. (A.23) and (A.24) above are related to the cosine and sine functions, respectively. In fact, for large N we have

$$\sum_{m=0}^{N-1} (-1)^m u_{2m}(x)\, u_{2m}\left(\sqrt{y_\nu^-/2}\right) = \frac{\cos\left(x\sqrt{y_\nu^-/2}\right)}{(2\pi)^{1/2}} + 0(N^{-1/2}) , \quad (A.25)$$

$$\sum_{m=1}^{N} (-1)^m u_{2m-1}(x)\, u_{2m-1}\left(\sqrt{y_\nu^+/2}\right) = \frac{\sin\left(x\sqrt{y_\nu^+/2}\right)}{(2\pi)^{1/2}} + 0(N^{-1/2}) . \quad (A.26)$$

Applying the asymptotic expressions (A.21)-(A.22) for $N_\nu^\mp$, the residues $R_\nu^\mp(x)$ are given, for large N, by

$$R_\nu^-(x) \sim \frac{\cos^2\left(x\sqrt{y_\nu^-/2}\right)}{2N^{1/2}} , \quad (A.27)$$

$$R_\nu^+(x) \sim \frac{\sin^2\left(x\sqrt{y_\nu^+/2}\right)}{2N^{1/2}} . \quad (A.28)$$

But from Eqs. (A.9) and (A.10)

$$\cos^2(x\sqrt{y_\nu^-/2}) + \sin^2(x\sqrt{y_\nu^+/2}) = 1 + 0(N^{-1}) \quad ,$$

and therefore, for large N

$$R_\nu^-(x) + R_\nu^+(x) \sim \frac{1}{2N^{1/2}} \quad . \tag{A.29}$$

The approximate integrated local density of states is given by Eq. (3.3). Thus, for large N we obtain from (A.29)

$$\begin{aligned} \mathcal{N}_N(x,E) &= \sum_{E^\nu \leq E} [R_\nu^-(x) + R_\nu^+(x)] \\ &= \frac{\nu_{max}}{2N^{1/2}} \quad , \end{aligned} \tag{A.30}$$

where ν_{max} is defined by

$$E^{\nu_{max}} = E$$

From Eqs. (A.9)-(A.10) ν_{max} is given by

$$\nu_{max} = 2(2EN)^{1/2}/\pi \quad . \tag{A.31}$$

Substituting this expression into Eq. (A.30) gives

$$\mathcal{N}_N(x,E) = (2E)^{1/2}/\pi \quad . \tag{A.32}$$

for large N, which is identical with the exact expression for $\mathcal{N}(x,E)$, Eq. (4.3).

REFERENCES

1. W. Kohn and W. H. Butler, J. Res. Nat. Bur. of Stand. Sect. A. **74A**, 443 (1970).
2. J. Keller, J. de Phys. Colloq. (Orsay, Fr.), Suppl. au N° 5-6, **33**, C3-241 (1972).
3. R. P. Messmer, S. K. Knudson, K. H. Johnson, J. B. Diamond, and C. Y. Yang, Phys. Rev. **B13**, 1396 (1976).
4. W. P. Menzel, K. Mednick, C. C. Lin, and C. F. Dorman, J. Chem. Phys. **63**, 4708 (1975).
5. F. Cyrot-Lackman, J. Phys. Chem. Solids **23**, 1235 (1968).

6. R. Haydock, V. Heine, and M. J. Kelly, J. Phys. **C8**, 2591 (1975).
7. W. Kohn and A. Yaniv, to be published.
8. P. Hohenberg and W. Kohn, Phys. Rev. **136B**, 864 (1964).
9. W. Kohn and L. J. Sham, Phys. Rev. **140A**, 1133 (1965).
10. F. M. Mueller, Phys. Rev. **153**, 659 (1967).
11. F. Perrot, private communication.
12. J. H. Wilkinson, *The Algebraic Eigenvalue Problem* (Clarendon Press, Oxford, 1965).
13. G. Szegö, *Orthogonal Polynomials* (American Mathematical Society, New York, 1959).

CERN'S FIRST DIRECTOR-GENERAL, 1954-1955

by L. Kowarski

ABSTRACT

Felix Bloch came to Geneva as the first Director-General of CERN in 1954. His role in shaping the early scientific structure and research atmosphere at CERN is described in this paper from a personal point of view.

Human achievements in all their variety are amenable to one universal criterion: that of novelty. The first achiever in any line of endeavor stands apart; after Edmund Hillary, no Everest-climber could claim to have done the same. Pioneering achievers show the way and merit our thanks; how much greater is the merit if the way not only has been trod for the first time, but is by itself the first of a new kind. Thus, when attempting to appraise the role of Felix Bloch in CERN's early years, we should keep in mind that there was no precedent to guide his first steps, that no organization directly similar to CERN was in sight for the purpose of a comparison. Bloch's value as an innovator was enhanced by the fact that the very existence of CERN[1] represented a major innovation in the history of scientific institutions.

The creators of CERN were called upon to innovate in three customarily distinct fields of human endeavor: scientific, engineering, and managerial. All three were essential. The scientists had to define the aim and the spirit of the whole enterprise; the engineers would handle its most conspicuous material aspects; the managers would set up and operate its institutions on an international scale. Ends and means had to be kept in balance: the holders of material and institutional powers might be tempted to indulge in self-justifying exercises in pure engineering or pure management. This possible danger was clearly perceived by CERN's scientific founders; ac-

L. Kowarski is deceased.

cordingly, they sought to strengthen the head of the scientific component by entrusting the top executive position in CERN's hierarchy to a creative scientist of a high level of achievement and recognition. A Nobel prize-winning physicist would be likely to fit this description; other cases and other criteria of outstanding value could of course be taken into consideration.

It was taken for granted, with a remarkable unanimity, that CERN *would* be headed by a chief executive officer. Remarkable, since this way of running an enterprise is by no means the only thinkable one. Supreme powers may be held in rotation, or collegially, or dually (this last variant is, in fact, the one now in force at CERN). Perhaps unity at the top appeared particularly desirable as a counterpart to disruptive effects of conflicting nationalisms. At any rate, speculations concerning various conspicuous "papabili" began to circulate as soon as CERN's governing Council started the discussion of the organization's future structure, early in 1953. My own memories of these rumors were appropriate to my standing in the hierarchy of the provisional CERN: close to the decision-making top but not of it, a high-level employee but not an employer. Three names percolated to us with some insistence: Casimir, Blackett, Auger; none of them fitted perfectly the already then current definitions of "the ideal man for the job," but then who ever does? The margin of acceptable variation is a matter of personal judgment; in this case Bohr's personality was preponderant enough to impose his idea of this margin. In making his choice he was supported by Heisenberg, whose enormous prestige on the highest level of European physics contributed a powerful influence in the same direction.

The Council appointed from its midst a "Nominating Committee," which started its work in the summer of 1953. By that time, Bohr's and Heisenberg's ideas were already quite clear, and the first contacts with Bloch had already taken place. Whether other contacts were attempted in parallel, and quickly discouraged, we do not know; perhaps the *a priori* definition of the set to be scanned was so narrow that the search for possible candidates was reduced to the question "are there any at all?" rather than "how many are there?" At any rate Bloch was a Nobel laureate, and a theoretician rather than an experimenter, a qualification we could expect to be significant in Bohr's and Heisenberg's eyes, and which legitimately could be claimed to guarantee a wider grasp of relevant general ideas. Bloch's enduring ties with his original background defined him as being "of" Europe, yet not "in" Europe, therefore not too closely identified with any particular geographic grouping. He was aware of the CERN initiative and had already followed sympathetically its development for several months before he became personally involved.

A long series of mutual questionings and "soul searching" (an expression Bloch used himself in a talk he had with me in 1974 on this subject) started then (both Bohr and Heisenberg got engaged in it practically at the same time), and went on, by correspondence, for almost a year before Bloch and Bohr had the opportunity to sum it up in a personal get-together (Copenhagen, Spring 1954).

The long exchanges that had led to this conclusion were conducted so discreetly that many of Bohr's colleagues on the Council were hardly aware of this vigorous attempt to shape CERN's future leadership in accordance with an explicit philosophy of scientific values. Other philosophies were not at all unthinkable at this stage: for example, the idea was widely held that CERN's whole effort during the first few years should concentrate on the construction and commissioning of the first major piece of equipment (the synchrocyclotron). The man who, since May 1952, was in charge of this particular sector (Professor C. J. Bakker of Amsterdam) was the logical choice for the temporary leadership on the lines so defined. Accordingly, in October 1953 the governing Council expressed an almost unanimous intention to nominate Bakker as its candidate for the post of Director-General, which was expected to be officially created some time in 1954. A strong dissent was, however, voiced by one member country—the Federal Republic of Germany, then represented by Heisenberg. Since for this all-important appointment unanimity was considered essential, the formal decision was postponed and, in fact, withdrawn from the agenda. Bohr's and Bloch's "soul searching" could then continue unhampered.

When they finally met, Heisenberg was not present, but his views were incorporated in Bohr's presentation. They reached, in outline, an agreement—not yet a final one (since many essential details had, as yet, to be left open), but firm enough to induce Bloch to get a closer look at the already existing concrete elements of the situation. He came to Geneva in that same spring of 1954.

He found there a very busy international community of some hundred members of the provisional staff; many of them were settled in Geneva since the last months of 1953, and their numbers were growing daily. They were engineers, technicians, office workers, and also a few physicists who were temporarily diverted from their "pure" research by the lure of taking part in a new departure, in the creation of new facilities that, once ready, would enable them to resume their fundamental pursuits. CERN was also carrying on its payroll a small number of physicists who were learning the techniques of particle physics as visitors in various European institutes. The design and procurement for the synchrocyclotron were at that time still located in Amsterdam. We usually called it the "small machine," reserving the name "big machine" to the Proton Synchrotron (PS), which was expected to take more time and become available a few years later than the "small" one. The design of the PS required preliminary experimentation on a small scale, and this was located in Geneva, first under F. Goward and later, after his untimely death in early 1954, under J. B. Adams. All surrounding "landscape" activities—administration, building, workshops, library services, were the task of the "Laboratory Group," which I headed with P. Preiswerk as my deputy. E. Amaldi, the Secretary-General of the provisional CERN, shared his time between his Rome office and his Geneva headquarters, where he ensured the detailed coordination between the groups listed above and also with the theoretical studies, then still located in Copenhagen.

Bloch's American experience still belonged to the traditional academic world; the Big Science spirit, already dominant at that time in neighboring Berkeley (the Radiation Laboratory), had not yet spread to Stanford. In Geneva, facing the busy beehive in the first stages of its self-affirmation, Bloch knew that the task before him required him first to preserve, and then to impose his own scale of values, making full use of the powers vested in his title and function. He would have to make decisions in a great variety of situations of which he would have first to grasp the unfamiliar elements. How much of his time, or peace of mind, would be left free so as to prevent the complete interruption of his own research activity? Since the spring of 1952 Amaldi had managed to direct the provisional CERN without having to relinquish his teaching and research duties in Rome, but during most of that period CERN was hardly more than an embryo. Would a fully grown CERN remain as accommodating? The first plunge gave Bloch a taste of the magnitude of the problem; a visit to Paris followed and there the issue was brought into the open.

Bloch's insistence on his continuing role in research physics was welcomed, and in fact asserted to be an essential component of a scientific personality deemed suitable for the post. He would be given a personal research budget and assistants of his choice would be put on CERN's payroll. The earmarking, for this purpose, of a substantial fraction of his time would require some statutory alleviation of the administrative load. A leading executive with fully deputized powers would have to be appointed, preferably a person of proven value both as a researcher and as a manager. It was considered probable that Amaldi would be available for such a position. The problem thus appeared to be neatly solved; the Nominating Committee was greatly relieved and, perhaps, in its eagerness, tended to minimize the gap between a mere probability and a firm prospect. Back in Geneva, when discussing these arrangements, I darkly muttered, "They are offering you a bribe." Many years later Bloch reminded me of this episode and seemed to agree when I pointed out that alluring bribes should, after all, be taken as compliments.

The expectation centered on Amaldi was, in fact, meant to be no less a compliment, but as the day of the binding decision was approaching, no conspicuous closing of the probability gap had taken place. At the end of

September 1954, the intergovernmental convention establishing CERN as a legal international entity came into force, and, a few days later, the Council in Geneva held its first session as the permanent CERN's governing body. Bloch arrived from America just in time to be officially confirmed as CERN's first Director-General. The appointment of his Deputy was expected to follow, but then it became known that Amaldi had not confirmed his availability. The Council proceeded to set up the divisional structure and to appoint the Division Heads. Three new divisions were created by expanding the preliminary nuclei that had originally been initiated within the "Laboratory Group"; of these three, the Division of Administration was given the widest powers and its Head, an experienced British civil servant (S. Dakin) became *de facto* Bloch's deputy.

Yet the desirable degree of blending between a scientific outlook and a decision-making managerial power could not be expected from Dakin and this responsibility had, therefore, to be assumed by Bloch himself. Thus came to nothing this ingenious compromise that, half a year before, appeared to be so necessary and so readily attainable. Within a few days of his arrival in Geneva Bloch had to face a clear situation: the position he had accepted did not, in fact, exist at all. It did not take him long to draw a conclusion that, to him, was obvious, no matter how surprising it appeared to those who did not expect him to stick to his sense of values so simply and wholeheartedly. He would give the new organization as much help as he could afford within the limits compatible with the continuation of his life as a working physicist. This meant that he had to remain on duty as long as (but no longer than) the time necessary for CERN to find and to appoint his successor and to ensure a reasonably smooth transmission of responsibility.

The prospective brevity of Bloch's tenure of office became widely known fairly soon; I remember discussing this subject with Dakin some six or seven weeks after his (and Bloch's) arrival in Geneva. On my own level of the hierarchy, as I remember it, the feelings were mixed. The prospect of continued uncertainty and instability at the top was viewed with misgivings; on the other hand, we understood and respected Bloch's motives. We appreciated his willingness to help, and we cooperated. Thus, during these months of transition and—for most of us—of settling down, the building up of CERN's equipment and institutions went on smoothly enough.

The same months of breathing space enabled the Council to seek a new solution of the top executive problem without undue haste. Arguments in favor of appointing Bakker, which had carried so much weight on a previous occasion (October 1953), could now be revived in a new context. Bloch's pioneering incumbency *did* assert the primacy of scientific values in the definition of the Director-General's function; a time-limited deviation from this definition, for the duration of a time-limited stage in CERN's life, could now more easily be accepted. With Bakker the machine-builder at the head of CERN, whatever remained to be done about the synchrocyclotron could now be safely left to Bakker's original team. The direct responsibility for the completion and commissioning of this machine would be given to an experimental physicist, whose hold on this (at that time) dominating piece of equipment would automatically put him in the position of CERN's Chief Scientist. The new key appointments were announced in the spring of 1955; Bloch's tenure terminated in September, and the new regime (Bakker as Director-General, W. Gentner as the Head of "SC," the Synchrocyclotron Division) entered into force with no delay or disturbance.

A new kind of organization, a new kind of task, the prospect of a very brief tenure, the early arrival of the successor-designate—how much opportunity for achievement can one find under such inauspicious conditions, how much of a lasting imprint can one hope to leave? What measure of success was attainable by proclaiming and adopting a certain philosophy of values even if the banner-bearer of these values did not remain available for any length of time? It is here that Bloch's initiatives and actions undertaken during these brief months of 1954-1955 are worth a closer look.

First, a word about the location. CERN was soon to be identified with a single piece of land, at the suburban village of Meyrin, but in 1955 the process of concentration was yet far from being completed. Some provisional premises were put at our disposal by the Geneva authorities, on a piece of ground directly adjoining the growing Geneva airport—a quaint old villa, a few corners in the maintenance hangars, a place to set up a few prefabricated wooden huts. The villa became the executive headquarters, where both Bloch and Dakin had their offices. The hangars housed our first workshops and such preliminary SC work as could already be undertaken in Geneva under the, as yet, remote authority first of Bakker, and later on of Gentner. Library, documentation, some apparatus design, and various other supporting activities took place in the huts, while a few kilometers to the west, at Meyrin, the bulk of the excavation and foundation work (with the SC Building as its first priority) was already in progress since May 1954. Another temporary focus of CERN activity had sprung up at the urban center of Geneva itself: the Physics Institute of the University played host to CERN-connected international gatherings (technical and managerial), to CERN's lectures and seminars and—last but not least—to some more CERN-owned prefabs. There the division in charge of the "Big Machine" (Proton Synchrotron), under J. Adams, was deploying its design offices and small-scale supporting experiments.

CERN's jurisdiction, CERN-managed financing and, therefore, some of Bloch's responsibilities still were directly connected with a few locations elsewhere in Europe. As we have already seen, the build-up of the scientific staff was initiated well in advance. These researchers were then still in temporary residence at various suitably equipped research centers—the future SC experimenters mostly in Liverpool and Uppsala, the future theoreticians mostly in Copenhagen. Cosmic-ray research, both experimental and theoretical, offered another set of possibilities for the initiation of CERN's future resident staff; by mid-1955, the attractive pull of Geneva was already bringing a team of cosmic-ray physicists, under A. Newth of Manchester, towards a nearby high-mountain location (at Jungfraujoch, well within a commuting radius from CERN's headquarters).

Bloch was fully aware of the relevance of these geographic aspects. Other leaders might feel the lure of a compromise that would enable them to combine a European faith with an unbroken allegiance to their own local fountainhead of scientific excellence. Could not a physicist based in Paris or Hamburg or elsewhere, just use the "Geneva machine station" as a source of data, as an oceanographer uses a dredging ship, or a geophysicist a balloon? Bloch allowed no such evasions. To promote the unity of European particle physics meant for him to foster this unity in its simplest spatial expression—one place in which the major equipment is located, the experiments are carried out, the abstract thinking goes on in daily contact with the experimenters. The machine-building and managerial anthill could safely be left to develop under its own momentum; *his* task was to see that a seed of scientific life was securely implanted in this newly prepared ground and that the young plant received proper care until it became self-sustaining.

The double life of Bloch as an executive and as a working physicist was reflected in the duality of his workaday arrangements. His directional offices at the airport provided the meeting ground for the frequent consultation sessions between the Division Leaders in which these (very autonomous) officers would tackle the current problems of common interest such as staffing, interdivisional exchanges of services and materials, etc. Contacts between the city-based PS Division and the rest of CERN were relatively few and hardly a source of major worries for the Director-General. The SC activities were closer at hand, and Bakker's gradual transition from the status of a Division Head to that of Director-designate brought him into increasingly closer contact with Bloch. In my own domain of "auxiliary science," the design and development of experimental equipment lay largely outside of Bloch's sphere of personal interest; only my library-and-publication section was for him an area of some attentive concern (see below). The building site at Meyrin was largely left in Preiswerk's care and Dakin was looking after its finances and contracts.

Conceived on these fairly reticent lines, it is hardly surprising that Bloch's executive personality left few outstanding traces in my memory. On one precise occasion—the selection and appointment of a Chief Purchasing Officer—Bloch chaired the Division Heads' meeting and expressed some of his opinions on the essence and psychology of administration, which I still remember with interest. My point here is that, having once accepted a task, no matter how uncongenial, he gave it his full attention and reflected on it in depth. He showed the same goodwill in carrying out those rare ceremonial duties that were inherent in high office and had to be attended to even at CERN, in spite of its resolutely matter-of-fact style. In June 1955, for some obscure reason, it was deemed appropriate to hold some sort of a "Foundation Stone" celebration; Bloch submitted to this ritual with a visible (and memorable) effort, yet with an untroubled good grace.

There were compensations. During a substantial fraction of his life in Geneva, Bloch was free, in his other personality, to pursue theoretical researches on nuclear magnetism. In parallel with his own leave of absence from Stanford, two of his American assistants (Drs. J. T. Arnold and W. A. Anderson), temporarily on the CERN payroll, continued to work under his supervision. The product of these associated efforts was, in fact, first published jointly as a CERN Report.

The two faces of Bloch's person at CERN—that of an organizational leader and that of a creative physicist—found their combined expression in one overriding preoccupation, in which Bohr's and Heisenberg's original sponsorship continued to be reflected. CERN was destined to become a focal location of advanced scientific research; it was therefore important to implant there, at the earliest possible stage, the customary hallmarks of a research community. There should be interdisciplinary contacts between various specialized sections; one of Bloch's first initiatives was, accordingly, the setting up of a periodical colloquium in which either resident or invited scientists would lecture on topics of general scientific interest. A central library can play a similarly unifying role; since library and documentation had remained under my jurisdiction, Bloch and I met there on a common ground of interest. He took the initiative of appointing an interdivisional Library Committee and saw in it another expression of the desired unity under CERN's scientific banner. He gave the same measure of his personal attention to CERN's beginnings in the role of a scientific publisher.

The most far-reaching manifestation of the same line of thought occurred in connection with the implantation of theoretical physics. Bloch insisted on the creation of a Geneva-located group of theoretical studies in particle physics and thus initiated a process of transfer from Copenhagen to Geneva that took several years to complete (pending the final installation of CERN at its Meyrin site). A corresponding concentration took place during the same years on the experimental side, where the synchrocyclotron, nearing its completion, was exerting an increasing attraction. No corresponding pull was felt by the theoreticians; there a conscious effort *had* to be applied, and there Bloch did fulfill that strengthening function his sponsors originally had in mind.

Looking back at the state and trends at CERN at the time of Bloch's departure, one can see that the supremacy of scientific values, which he had served so wholeheartedly during his brief tenure, was accepted unquestioningly and finally. It is difficult to imagine today that it could have been otherwise; yet the history of human institutions often shows that aims and intentions do not necessarily coincide with those that had animated the founders. When they do, this may be due to a strong guiding hand initially at work. Bloch's immediate successors—Bakker as Director-General, Gentner as the de facto Chief Scientist who presided over CERN's beginnings in experimental physics—continued in accordance with the same principles. The process of building up CERN's major facilities went on into the early 1960s. By that time a new structure became necessary. (It was essentially provided by Adams during his brief passage as Director-General.) In its new maturity CERN was then ready to experience another period of putting the supreme executive power in the hands of a renowned master of creative physics. Under V. F. Weisskopf's leadership CERN asserted its place in Europe and in the world and thus brought to fulfillment those same expectations that originally determined Bohr's and Heisenberg's choice of Felix Bloch as CERN's first Director-General.

REFERENCE

1. A brief description of CERN is given in the Editor's Note to the article by L. Kowarski, Bull. At. Sci. **33**, #7, 44 (1977).

FIVE ENCOUNTERS WITH FELIX BLOCH

by Willis Lamb

ABSTRACT

The impact of Felix Bloch's work on the fields of parity non-conservation, the Mössbauer effect, nuclear induction, chemical shifts, and laser theory is described from a personal point of view.

I have benefited enormously from contacts with a number of the great theoretical physicists of the twentieth century. One such relationship far exceeds all the others in respect to duration and meaning to me. I am going to describe five encounters with Felix Bloch that involved purely scientific matters. Personal matters come into the account only to set the times and places of the episodes related. The first two and the last of the five encounters involved suggestions Felix made to me in connection with my research. If I had had the wit, energy, and luck to follow up properly the earliest two of these leads, I might have made some very good discoveries. The third encounter involved a request from him for help on his research. It turned out that he did not, after the fact, need this help. Still, the story has a certain interest for me and might provide a footnote for a history of modern physics.

The fourth encounter was very slight. I had worked on a certain problem in the early forties. Bloch was working in the same area in the early fifties. I had missed following up some interesting aspects of my work. Bloch and I talked about his problem and its relationship to my earlier research. It later turned out that his work was capable of enormous application, in chemistry rather than in physics, so that he did not follow it up as far as he might have done.

Willis Lamb is Professor of Physics at the University of Arizona.

The fifth encounter led me to the development of the theory of lasers and masers, which has been a very productive and useful field of research for over twenty years.

I received a B.Sc. at Berkeley in chemistry in 1934, and began the study of theoretical physics with Robert Oppenheimer. In those days, the Physics Departments at Berkeley and Stanford occasionally had joint picnics followed by a colloquium. One that I remember vividly took place in the spring of 1934 at the Lick Observatory on Mount Hamilton. The speaker was someone from Stanford named Felix Bloch. Solid state physics was not taught at Berkeley then, and I had no notion of the enormous contributions made to that field by the speaker. He had just come from Rome and told us about Fermi's new work on the theory of beta decay.

In the fall of 1934, Oppenheimer and Bloch began to conduct joint weekly theoretical physics seminars alternating between Berkeley and Stanford. During the next year I gradually got to know Bloch, who accepted me with an unusual degree of tolerance. In the summer of 1935, I went to the theoretical physics school at Ann Arbor. This was organized by George Uhlenbeck and Samuel Goudsmit and operated during the eleven years before the U.S. entry into World War II. Bloch was one of the lecturers that summer, and I learned from him the elements of his theory of electrical conductivity. I also acquired a little understanding of the theory of beta decay from other lecturers at the summer school, among whom were Enrico Fermi and David Dennison, as well as Uhlenbeck and Goudsmit.

PARITY NON-CONSERVATION

One day in the autumn of 1935, I was driving with Bloch somewhere in Berkeley, and he told me that the shape of a beta decay electron energy spectrum would depend on the change of angular momentum of the nuclear states. He suggested that I calculate this effect in more detail.

The angular momentum dependence contemplated by Bloch came from the variation across the nuclear radius of the plane wave electron and neutrino wave functions in the transition matrix element. This brought in additional momentum-dependent factors and thereby changed the shape of the theoretical electron energy spectrum.

At that time, the theory of beta decay was very primitive and poorly formulated. Competing with Fermi's originally postulated interaction for the emission of an electron and a neutrino was the proposal of Konopinski and Uhlenbeck that explicitly involved derivatives of lepton wavefunctions, and also changed the momentum dependence of the matrix element for the emission process. Gamov and Teller had also suggested an interaction that differed from Fermi's by the presence of nucleon and lepton spin operators.

By that time, I had learned from Pauli's 1933 *Handbuch der Physik* article[1] on Quantum Mechanics how he had classified the covariant quantities that could be made out of binary combinations of Dirac wave functions: scalar, polar vector, tensor, axial vector, and pseudovector. A particular combination of these interactions was used by Fermi, and, stimulated by Bloch, I began to investigate the effect on the beta spectrum of taking wild mixtures of the various Pauli building blocks. I remember that Robert Serber (then at Berkeley) showed me that some of the mixtures I was considering did not conserve parity! This might have been responsible for a missed opportunity, but 1936 was simply too early for the discovery[2] of non-parity conserving interactions. In fact, it took over twenty years before physics was ready for the theoretical analysis of Lee and Yang.

The experimental state of measurement of beta spectra in 1936 was very confused because of thick source effects. I never really accomplished anything very useful with beta decay theory, and the work was published[3,4] only in two abstracts of talks given before meetings of the Pacific Coast Section of the American Physical Society.

SUMMERS AT STANFORD

There wasn't much theoretical physics going on in Berkeley during the summers. Oppenheimer invariably spent about six weeks in the late spring at Cal Tech. I usually followed him there, but never to his summer retreat in New Mexico. Bloch stayed working at Stanford during the summers, and kindly provided me with a desk and library facilities. I was there during the summers of 1936–1939 and 1941. Each year there was a visiting lecturer, and in these years I came to know George Gamov, Edward Teller, Victor Weisskopf, J. H. Van Vleck, George Placzek, Samuel Allison, I. I. Rabi, George Kistiakowski, and many others who passed through Palo Alto on their summer travels. I also met most of the Stanford physicists: David Webster, Paul Kirkpatrick, Norris Bradbury, William Hansen, Russell Varian, Arnold Siegert, and Arnold Nordsieck (I had already known Nordsieck at Berkeley, and we later were colleagues at Columbia for many years). My meeting with Rabi led to an appointment as Instructor in Physics at Columbia in the fall of 1938 after I received a Ph.D.

THE MÖSSBAUER EFFECT

At Ann Arbor, I had heard Fermi lecture on the effect of chemical binding of a hydrogen atom on its scattering of slow neutrons. This interested me, and I began to work on related problems. It seemed that there might also be an effect of the binding of a hydrogen atom on the capture cross section for slow neutrons. At first I thought the effect would be large, but finally had to settle for a very rough estimate of the cross section for a very unlikely process: the radiationless capture of neutrons by bound protons to form deuterons, with the excess energy and momentum going into vibrational motion of the deuteron instead of a gamma ray. The normal capture process was very little affected by the chemical binding. Even today, this radiationless capture has never been seen, but I am still hoping that someday it may be. This work[5] formed part of my doctoral thesis. The other part[6] dealt with electromagnetic properties of nuclear matter.

That summer, I came across a paper by Hans Bethe and Placzek[7] on the resonant capture of slow neutrons by nuclei in a gas. The resulting compound nucleus could decay in a number of ways that contributed to the natural width of the resonance curve. The Lorentzian resonance curve of the Breit-Wigner theory of capture of slow neutrons to form a compound nucleus was modified because of the Doppler effect. In the limit of small gas atom velocities, the resonance curve was Lorentzian, while for rapid motions it became a Doppler-broadened Gaussian. In the general case, Bethe and Placzek obtained the Voigt[8] profile, which was well known in the theory of Doppler broadening of naturally broadened spectral lines emitted or absorbed by atoms in a gas.

Since most of the experimental studies on resonant neutron capture involved atoms in a solid, I began to generalize the Bethe-Placzek theory to deal with nuclei bound in a crystal lattice. For the case of a simple harmonic oscillator (Einstein) model of the solid, it was clear that the neutron resonance curve would be a sum of Lorentzians with centers displaced by the separation of the vibrational energy levels of the oscillator. Then using what was learned from Bloch in a previous summer, I worked out the theory for a Debye model solid. Instead of a single oscillator frequency, there was a frequency continuum extending from zero up to a maximum value simply related to the Debye temperature of the solid. I derived a general, but complicated, expression for the neutron capture cross section as a function of neutron energy. The equations are reproduced here:

$$W(E) = \frac{2}{\Gamma} \text{ Real} \int_0^\infty d\mu \, \exp\left[i\mu(E-E_o) + \frac{i\Gamma}{2} + g(\mu)\right] \quad .$$

E is the neutron energy, E_o its resonance value without allowance for recoil, and Γ is the resonance width. The function $g(\mu)$ is given by a sum

$$g(\mu) = \sum_s q_s^2 \left\{ (\overline{\alpha}_s + 1)e^{-i\mu\hbar\omega_s} + \overline{\alpha}_s e^{i\mu\hbar\omega_s} - 1 - 2\overline{\alpha}_s \right\}$$

over the normal modes s of the crystal lattice. The quantity $\overline{\alpha}_s$ is the thermal average number of phonons in mode s

$$\overline{\alpha}_s = \frac{1}{(e^{\hbar\omega_s/kT} - 1)}$$

where ω_s is the circular frequency and T the absolute temperature.

The dimensionless quantity

$$q_s^2 = \frac{(\vec{\mathbf{p}}\cdot\hat{\mathbf{e}})^2}{2NM\hbar\omega_s}$$

is essentially the ratio of a recoil energy to the phonon energy $\hbar\omega_s$. The recoil energy is that given to the whole crystal of N atoms of mass M by the capture of a neutron of momentum $\mathbf{p}$. For large N, the sum over modes can be replaced by an integration, and N drops out of the final answer.

In the limit of low Debye temperature, the equations can be greatly simplified, and they then yield the Voigt profile of the Bethe-Placzek theory. In a gas, the capture of a neutron leads to a recoil of the compound nucleus. The resonance energy is shifted from its ideal value by the amount R of the recoil energy. In the limit of very high Debye temperature (stiff binding), the resonance is a Lorentzian function of the expected natural width, but without recoil shift. The momentum of the neutron, but no appreciable kinetic energy, is transferred to the whole crystal lattice.

I wrote a paper describing this work and had valuable suggestions from Nordsieck, Teller, and Placzek. Then I showed it to Bloch. He made the first real criticism of the paper: the results were contained in the complicated equations given above. The usefulness of the work could be greatly increased if it were shown that the equation could be used to produce a graph. I attempted to persuade Bloch that I had obtained the interesting limiting cases analytically, but to no avail. He wanted a figure of a non-trivial intermediate case. When I went to Columbia my first research involved many days at an electric Marchant calculator. This was very tedious, but finally I had evaluated the necessary multiple integrals, and prepared some crude hand-drawn figures so that the paper[9] could be sent in for publication. One of these figures is shown here (figure 1).

I also sent in an abstract of the work[10] for the Thanksgiving 1938 meeting of the American Physical Society. I no longer remember what kept me from going to the meeting, but I had to look around for a colleague willing to give the presentation in my place. Julian Schwinger, who was a Columbia

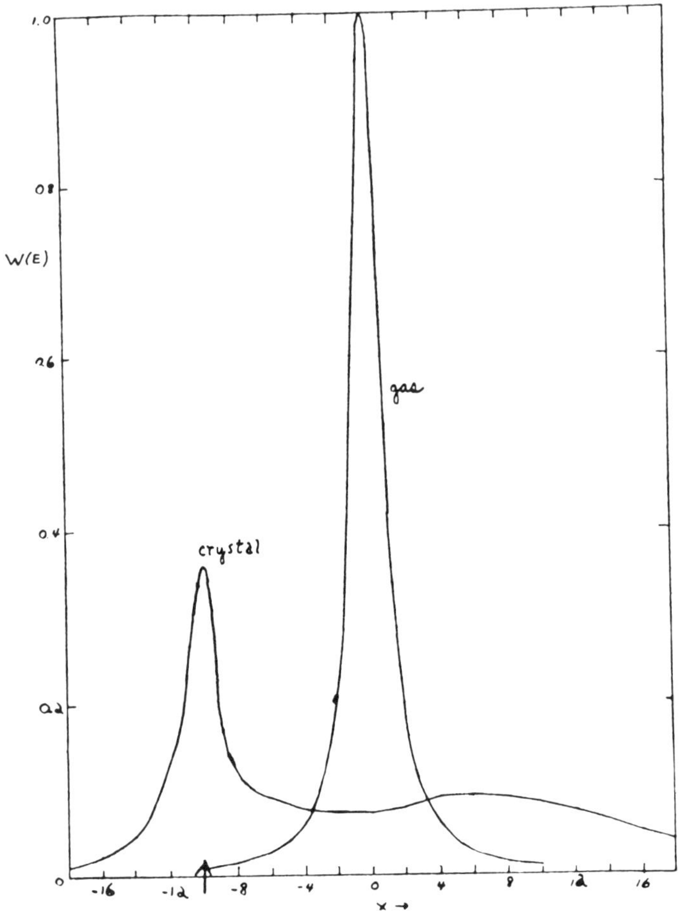

FIG. 1. PLOT OF THE NEUTRON RESONANCE ABSORPTION CURVE in cold solid silver for an assumed value of line width equal to one quarter of the Debye energy. The curve one would obtain with free atoms is shown for comparison. The abscissa measures the energy distance from resonance in units of the half width. If the lattice binding were very strong, the curve for the crystal would have the same form for the gas, except that it would be centered about the point shown by the arrow.

graduate student at that time, was scheduled to talk at Chicago on his work[11] on the quadrupole moment of the deuteron. I asked him to substitute for me, and after glancing at the paper he agreed to do so. I never got any fan mail from his talk, but had considerable pleasure in later years when Julian told me that he had learned some useful tricks from the paper, based on the fact that in a problem with a continuum of modes, it is very unlikely that two or more phonons are emitted into the same mode.

My 1938 work applied to capture of neutrons in a crystal. Some experimental confirmation of the calculations was obtained[12] at Brookhaven National Laboratory. However, the observable effects were in practice relatively small because the nuclear recoil energy was not very large compared to the resonance width. It should have been obvious to me that the analysis given above would apply as well to absorption of gamma rays as to the capture of neutrons. Ordinarily the recoil energy of a nucleus during absorption of a gamma ray would be small compared to the resonance line width. In the case of transitions involving long-lived nuclear isomeric states, however, the natural width could be exceedingly small. Of course, in 1938, I had never heard of isomers. In later years many people were looking for resonant absorption in target nuclei of gamma radiation that had been emitted by the same kind of nuclei in a source. The difficulty of observation came from the recoil losses of energy on emission and absorption, together with the very narrow resonant line width. Several ingenious methods were devised to make up for the recoil energies of the emission and absorption processes. The first success in this field was achieved by Philip Moon[13] of the University of Birmingham. I met him in England shortly after moving to Oxford in 1956, and he said that we should have a talk about the implications of my 1939 paper for his work. Unfortunately, this invitation was not followed up by a meeting. Within a year, I received two reprints from Rudolf Mössbauer[14] of Heidelberg and Munich that told of his discovery and indicated that he had been referred to my old paper by one of his teachers, Hans Jensen. The graphs Bloch had persuaded me to publish gave Mössbauer the clue he needed to understand the nature of his effect.

The existence of the Mössbauer effect provides a very neat proof[15] that radiation is emitted "instantaneously." When a nuclear isomer with a lifetime of perhaps a microsecond emits gamma radiation, the emission time might be uncertain by several lifetimes, but the time taken for the actual emission process must be short compared to the sub-picosecond periods of the lattice vibrations. Otherwise, the radiative recoil of the nucleus would be stretched out over several vibrational periods, and the vibrational quantum numbers of the nucleus would not change with any appreciable probability during the nearly adiabatic emission process. (Actually, a theoretical analysis shows that the time taken is of the order of the much shorter time it takes light to traverse the nuclear radius.)

The Mössbauer emission and absorption of radiation can be thought of as simply giving a sudden impulse-like blow to the nucleus in the host lattice. This provides a very helpful picture for many important features of the Mössbauer and other radiative effects.

It was only after such considerations that I realized there was not much difference between the theory of the incoherent scattering of x rays (Debye-Waller theory) and the theory of the Mössbauer effect. The scattering of an x ray also transfers a momentum to an atom that can be replaced by an impulsive force. (Bloch did not tell me that, although he must have known it.)

NUCLEAR MAGNETIC INDUCTION (OR RESONANCE)

During the war years 1943–1945 I worked at Columbia on microwave magnetron oscillators. In 1945 I had occasion to visit Bloch at Stanford, and he made a request that I found slightly mysterious, but was happy to grant. He wanted to have a permanent magnet of the sort used to provide magnetic fields for magnetrons. I explained that our magnetrons were rather small, and since the field had to be high, we had to attach to the usual horseshoe magnet some tapered pole pieces of high permeability iron. These increased the magnetic field, but considerably reduced the spatial extent of the strong field region. Felix wanted a larger region of strong field, so we agreed that I would send him a horseshoe magnet with pole tips removed, magnetized to the highest degree possible with our facilities. On return to New York, I hastened to get a spare magnet and to magnetize it as strongly as I could. I then sent this to Bloch by parcel post.

Even after the wonderful discovery of nuclear induction in 1946 I had no idea why he had wanted the magnet. He told me the reason some years later. The conventional wisdom in 1945 was that the time required for attainment of thermodynamic equilibrium of a system of nuclear magnetic moments in a solid was very likely quite long. In fact, Bloch and I had both heard Van Vleck lecture in the summer of 1941 on the theory that clearly seemed to indicate a very long relaxation time. In 1945, Felix was thinking of the forthcoming attempt to detect nuclear magnetism. To do this he needed to have the spins partially oriented along a magnetic field. Thinking that this might take a long time, he wanted to have a sample containing protons "soaking" in a magnetic field for a time sufficiently long for thermal equilibration so that there could be a net nuclear magnetism. As the work proceeded, Bloch realized that doping a liquid sample by paramagnetic impurities would work much better. Hence the sample of protons soaking for many months in my magnet was never used. Still, it might have been necessary. Bloch never offered to return the magnet. He did, however, say that he did not think it was as strongly magnetized as I had measured it to be. I suppose the postal authorities may have opened the package and partially demagnetized the strange steel object. Bloch did not mention this episode in his Nobel Lecture of 1952.[16] Perhaps he thought it would embarrass me!

CHEMICAL SHIFTS

About 1939, it became a matter of importance to members of the molecular beam laboratory at Columbia to know the precise value of the magnetic field acting on a nucleus in an atom or molecule. The largest part of this is the externally applied field. However, the nearby electrons acquired an induced diamagnetic current, which produced an Amperian contribution to the magnetic field at the nucleus amounting to about 10 ppm of the applied field. This was roughly estimated by Sydney Millman and Polykarp Kusch and Rabi.[17] I made a more quantitative calculation for the case of atoms. I could solve the problem exactly for a hydrogen atom, and quite well, using the Fermi-Thomas model, for more complicated atoms. I did not try to do anything with molecules. My work was published under the slightly cryptic title "Internal Diamagnetic fields."[18] Later, Norman Ramsey showed how to make the corresponding calculation[19] for the hydrogen molecule.

In 1951, I joined the Stanford Physics Department faculty. At that time, Bloch was just discovering that super-high-resolution nuclear magnetic resonance made it possible to resolve the resonances[20] given by the magnetic moments of protons in different parts of a hydrogen-bearing organic molecule. These effects were clearly caused by internal diamagnetic fields, but were much easier to see than they had been in the earlier molecular beam experiments. The chemical shifts have turned out to be one of the most important tools for work in structural chemistry.[21] Bloch applied the method to some simple hydrocarbons, but as a physicist he did not wish to get involved with more complicated molecules. In subsequent years, John Pople[22] and others gave fruitful theoretical methods of analysis of the internal diamagnetic fields for complicated molecules. With the general commercial availability of super-high-resolution NMR apparatus, the method of chemical shifts has attained very wide and fruitful application. Certainly it is within the range of plausible possibilities that Bloch might receive a richly deserved Nobel Prize in Chemistry for his pioneering work in this field. (He also could easily have received a Physics prize for his work on electrical conductivity,[23] ferromagnetism,[24] magnetic scattering of neutrons,[25] or the first precise measurement of the magnetic moment of the neutron.[26])

LASER THEORY

The theory of an idealized two-level system has played a very important role in the analysis of all kinds of resonance physics. The theory was developed in the late thirties in independent papers by Schwinger,[27] Rabi,[28] and Bloch.[29] I used these techniques in the analysis of the hydrogen fine structure work with Retherford.[30] Following Bethe,[31] the radiative decay of the $2p$ level was simulated by adding to one of the equations a term with a damping constant.

At Stanford, I worked with Theodore Maiman[32,33] and with Irving Wieder[34] on the triplet fine structures of the $n = 2$ and $n = 3$ states of the helium atom. The two-level theory was still useful. Decay constants had to appear in each equation, but since they were equal, the algebra was simple.

On the other hand, my work with Michael Sanders and with Lee Wilcox[35] on the fine structure of the $n = 3$ states of hydrogen involved two (and more) level problems in which the states involved had different decay constants. It was still possible to insert radiative damping terms into the equations, but the algebra became enormously more complicated for the analysis of the $n = 3$ resonances. One such calculation took me ten pages of lengthy equations.

The way out of this difficulty was pointed out to me by Bloch. I was trying to solve a transient problem, but the measurements involved a steady state of systems excited by electron bombardment and decaying through the emission of radiation. Felix showed me how the use of the density matrix techniques could lead in about one-half page to the same result I had obtained with the wave equations linear in the probability amplitudes. The density matrix equations were bilinear in the probability amplitudes, more complicated and more numerous than the amplitude equations. The observed resonance signal was bilinear in probability amplitudes, however, and hence more readily calculated using the density matrix techniques. A very important simplification came by looking for quasi-steady state solutions of the density matrix equations instead of transient solutions of the time dependent wave equations.

I later found this technique of great value for formulating my theory[36] of a laser. Density matrix methods are almost universally used now, but very few people used them in 1955 when Bloch taught them to me. Of course, I had read papers[37] by Dirac and by von Neumann, who used density matrices. The work of Landau was not then known to me. I had not been clever enough to see the potential usefulness of the density matrix, and was enormously impressed by what I assumed to be Bloch's breadth of scholarship and deep physical insight.

A few years later, however, I learned that although Bloch had known of density matrices from the 1930 era, their utility in the fifties was suggested to him by Leonard Schiff. The first and second edition of Schiff's textbook on quantum mechanics (1939, 1955) did not have a single word about density matrices. When Schiff was asked why not, he is reported to have replied something like, "I was writing a text book on quantum mechanics and I thought that density matrices were a part of statistical mechanics." The third edition, which appeared in 1969, did have a brief section dealing with density matrices.

After hearing of these things, I remembered a conversation with Llewellyn Thomas at the Columbia Faculty Club in the late forties. Thomas was teaching quantum mechanics that semester. This was a course I had taught previously, and I asked him how he dealt with the subject. He replied, "I begin with the density matrix." Naturally, I wanted to know why he proceeded in such an unconventional fashion. He explained that he wanted to discuss the most general state before he considered special states. I have learned a great deal from Thomas, but usually it took me several years to understand what he told me. Incidentally, Schiff had learned his early quantum mechanics from Thomas at Ohio State.

IN CONCLUSION

The story of these encounters with Felix Bloch is necessarily very sketchy and incomplete. The five episodes were immensely valuable to me, even if I missed most of the opportunities he gave me. The reader can take it for granted they are only part of the story of my entire relationship to Felix, and that my words could never express how much this has meant to me.

REFERENCES

1. W. Pauli, "Die allgemeinen Prinzipien der Wellenmechanik," *Handbuch der Physik* (Springer, Berlin, 1933), vol. 24/1, esp. pp. 214-227.
2. T. D. Lee and C. N. Yang, "Parity nonconservation and a two component theory of the neutrino," Phys. Rev. **105**, 1671-1675 (1957).
3. W. E. Lamb, Jr., "On the unobservable decay of sodium^{22}," Phys. Rev. **50**, 388(A) (1936).
4. W. E. Lamb, Jr., "On the continuous beta spectrum of phosphorus^{32}," Phys. Rev. **51**, 145(A) (1937).
5. W. E. Lamb, Jr., "A note on the capture of slow neutrons in hydrogenous substances," Phys. Rev. **51**, 187-190 (1937).
6. W. E. Lamb, Jr. and L. I. Schiff, "On the electromagnetic properties of nuclear systems," Phys. Rev. **53**, 651-661 (1938).
7. H. A. Bethe and G. Placzek, "Resonance effects in nuclear processes," Phys. Rev. **51**, 450-484 (1937).
8. See M. Born, *Optik* (Springer, Berlin, 1933), pp. 482-486.
9. W. E. Lamb, Jr., "Capture of neutrons by atoms in a crystal lattice," Phys. Rev. **55**, 190-197 (1939).
10. W. E. Lamb, Jr., "Capture of neutrons by atoms in a crystal lattice," Phys. Rev. **55**, 234(A) (1939).
11. J. Schwinger, "On the neutron-proton interaction," Phys. Rev. **55**, 235(A) (1939).
12. V. L. Sailor, Brookhaven National Laboratory Report BNL-257 (T-40) 1953; Phys. Rev. **91**, 53 (1953); H. H. Landon, Phys. Rev. **93**, 931(A) (1954).
13. P. B. Moon, "Resonant nuclear scattering of gamma rays: theory and preliminary experiments," Proc. Phys. Soc. London, Sect. A **64**, 76-82 (1951). Other methods are described by K. G. Malmfors, "Resonant scattering of gamma rays," in K. Siegbahn, *Beta and Gamma Ray Spectroscopy* (Academic Press, New York, 1955), pp. 521-530.
14. R. L. Mössbauer, "Kernresonanzabsorption von Gammastrahlung in Ir^{191}," Naturwissenschaften **45**, 538-539 (1958); "Kernresonanzfluoreszenz von Gammastrahlung in Ir^{191}," Z. Phys. **151**, 124-143 (1958).

15. W. Lamb, Jr., "Panel discussion: theory," in *The Mössbauer Effect*, ed. by D. M. Compton and A. H. Schoen (Wiley, New York, 1962), pp. 17-35.
16. F. Bloch, "The principles of nuclear induction," *Le Prix Nobel en 1952* (Norstedt, Stockholm, 1953), pp. 83-96.
17. P. Kusch, S. Millman, and I. I. Rabi, "Nuclear magnetic moments of nitrogen^{14}, sodium^{23}, potassium^{39} and caesium^{133}," Phys. Rev. **55**, 1176-1181 (1939).
18. W. E. Lamb, Jr., "Internal diamagnetic fields," Phys. Rev. **60**, 817-819 (1941).
19. N. F. Ramsey, Jr., "Magnetic shielding of nuclei in molecules," Phys. Rev. **78**, 699-703 (1950); *Molecular Beams* (Oxford, 1956), pp. 162-166. See also A. Abragam, *The Principles of Nuclear Magnetism* (Oxford, 1961), pp. 175-191.
20. F. Bloch, "Nuclear magnetism," Am. Sci. **43**, 48-62 (1955). See also W. G. Proctor and F. C. Yu, Phys. Rev. **77**, 716-717 (L) (1950); W. C. Dickinson, Phys. Rev. **77**, 736-737 (1950); G. Lindstrom, Phys. Rev. **78**, 817-818 (1950); J. T. Arnold, S. S. Dharmatti, J. Chem. Phys. **19**, 507 (1951).
21. This could be seen by examining issues of such magazines as the Journal of Magnetic Resonance, 1968— (Academic Press, N. Y.); Organic Magnetic Resonance, 1969— (Heyden, London); the series, Progress in Nuclear Magnetic Resonance Spectroscopy, 1966— (Pergamon, Oxford); and extensive note books of *Sadtler's Standard N. M. R. Spectra*, 1976— (Sadtler, Philadelphia), and of the Thermodynamics Research Center, 1975—, Texas A&M University, sponsored by the American Petroleum Institute. See also J. Memory, *Quantum theory of magnetic resonance parameters* (McGraw-Hill, New York, 1968).
22. For example, J. A. Pople, "Chemical shifts in nuclear magnetic resonance," Proc. Roy. Soc. London **239**, 541-549 (1957).
23. F. Bloch, "Uber die Quantenmechanik der Electronen in Kristallgittern," Z. Phys. **52**, 555-600 (1928); "Zum elektrischen Widerstand bei tiefen Temperaturen," Z. Phys. **59**, 208-214 (1930).
24. F. Bloch, "Molekular Theorie des Magnetismus," in E. Marx's *Handbuch der Radiologie* (Akademische Verlag, Leipzig, 1934), vol. 6, part 2, pp. 375-484. Also F. Bloch, Z. Phys. **57**, 545-555 (1929); **61**, 206-219 (1930); **74**, 295-335 (1932); F. Bloch and G. Gentile, Z. Phys. **70**, 395-408 (1931).
25. F. Bloch, "Magnetic scattering of neutrons," Phys. Rev. **50**, 259(A) (1936); **51**, 994(L) (1937).
26. L. W. Alvarez and F. Bloch, "A quantitative determination of the neutron moment in absolute nuclear magnetons," Phys. Rev. **57**, 111-122 (1940).
27. J. Schwinger, "On nonadiabatic processes in inhomogeneous fields," Phys. Rev. **51**, 648-651 (1936).
28. I. I. Rabi, "Space quantization in a gyrating magnetic field," Phys. Rev. **51**, 652-654 (1936).
29. F. Bloch, "The magnetic moment of the neutron," Ann. inst. Henri Poincaré **8**, 63-78 (1938); see F. Bloch and A. J. Siegert, "Magnetic resonance for non-rotating fields," Phys. Rev. **57**, 522-527 (1940).
30. W. E. Lamb, Jr., "Fine Structure of the hydrogen atom, III," Phys. Rev. **85**, 259-276 (1952).
31. H. A. Bethe, "Quantenmechanik der Ein- und Zwei-Elektronen Probleme," *Handbuch der Physik* (Springer, Berlin, 1933), vol. 24/1 esp. pp. 273-560.
32. W. E. Lamb, Jr., "Microwave technique for determining the fine structure of the helium atom," Phys. Rev. **105**, 559-572 (1957).
33. W. E. Lamb, Jr., and T. H. Maiman, "Measurement of the fine structure separation $3^3P_1 - 3^3P_2$ for the helium atom," Phys. Rev. **105**, 573-579 (1957).
34. W. E. Lamb, Jr., and I. Wieder, "Fine structure of the 2^3P and 3^3P states in helium," Phys. Rev. **107**, 125-134 (1957).

35. W. E. Lamb, Jr., and T. M. Sanders, Jr., "Fine structure of $n = 3$ hydrogen by a radio-frequency method," Phys. Rev. **103**, 313-314 (1956). Also, "Fine structure of short-lived states of hydrogen by a microwave-optical method," Part I, Phys. Rev. **119**, 1901-1914 (1960), and Part II, W. E. Lamb, Jr., and L. R. Wilcox, Phys. Rev. **119**, 1915-1933 (1960).
36. W. E. Lamb, Jr., "Theory of an Optical Maser," Phys. Rev. **134A**, 1429-1450 (1964). Compare with W. E. Lamb, Jr., "Quantum Mechanical Amplifiers," *Lectures in Theoretical Physics*, vol. II, ed. by W. E. Brittin and B. W. Downes (Interscience, New York, 1960), pp. 435-483. See also M. Sargent, III, M. O. Scully, and W. E. Lamb, Jr., *Laser Physics* (Addison-Wesley, Reading, Mass., 1974, 1977).
37. For a good general reference on density matrices, see D. ter Haar, Rep. Prog. Phys. **24**, 304-362 (1961). Closer to laser theory are my "Applications des techniques de matrice densité à quelques problèmes de double resonance," J. Phys. Radium **19**, 801 (1958), and "Notes on the use of density matrices in quantum electronics," in *Laser Photochemistry, Tunable Lasers and other Topics*, ed. by S. F. Jacobs et al. (Addison-Wesley, Reading, Mass., 1976), pp. 403-420.

SOME ASPECTS OF NATURAL-ABUNDANCE NITROGEN-15 NUCLEAR MAGNETIC RESONANCE SPECTROSCOPY

by John D. Roberts

ABSTRACT

Some of the problems involved in obtaining useful nuclear magnetic resonance signals from nitrogen-15 at the natural-abundance level in organic compounds are discussed. Special attention is given to the structural, stereochemical, and solvent effects on the ^{15}N NMR spectra of amines. Applications of ^{15}N NMR spectra to structural analysis of sulfonamides, oximes, imines, and peptides are presented along with procedures for determining the exchange rates of N-H protons through acid or base catalysis.

Nuclear magnetic resonance (NMR) spectroscopy has expanded like the cosmic "big bang" during the thirty years following the initial developments by Bloch[1] and Purcell,[2] and the end seems nowhere in sight. No other form of spectroscopy has so many applications or provides such deep insight into structural problems of substances as liquids or in solutions. Other than quantum mechanics, there has been no greater gift of modern physics to chemistry.

The expansion of NMR in chemistry has hardly been uniform, because the technology for many possible applications has developed unevenly. As each new breakthrough has come about, however, exploitation of it has usually been very rapid indeed. Cases in point are ^{13}C and ^{15}N NMR spectroscopy at the low natural-abundance levels of these isotopes. The pioneering work of Lauterbur[3] on ^{13}C NMR (1957–1961) and that of Grant[4] starting in 1964 clearly established the potential of ^{13}C NMR for structural analysis, but the difficulty in taking high-resolution spectra at the natural-

John Roberts is Institute Professor of Chemistry at the California Institute of Technology.

abundance level, where the sensitivity is 1.76×10^{-3} that of ^{1}H, was not overcome until a series of technological improvements culminated in 1966–1968. These included 1) greater magnetic field homogeneity, which allowed use of larger samples; 2) more stable magnetic fields and RF oscillators, which allowed time averaging of weak signals; and 3) broad-band proton decoupling, which greatly increased the signal strength through collapsing the C-H spin-spin splittings and generation of a very favorable nuclear Overhauser enhancement. Subsequent application of the pulsed Fourier transform (FT) NMR technique supplied the final impetus for conversion of natural-abundance ^{13}C NMR into an essentially routine (albeit expensive) procedure for structure analysis that is now in very wide-spread use.[5,6,7]

Nitrogen NMR has a little different history. The chemical shift was in fact discovered by Proctor and Yu[8] for ammonium nitrate and much effort has been expended on developing ^{14}N NMR as a structural tool, principally by Klein, Richards, Witanowski, and their coworkers.[9] Although the range of nitrogen NMR chemical shifts is very large, however, the substantial quadrupole moment of the ^{14}N nucleus in conjunction with substantial electrical field gradients at nitrogen nuclei usually causes rapid relaxation and very substantial line broadening.[10] These line broadenings may be as large as 200 ppm and hence may completely obscure the fine details of the spectrum, especially spin-spin couplings, which with ^{14}N are usually less than 100 Hz. Conspicuous exceptions are provided by isocyanides and compounds that have a high degree of electrical symmetry around nitrogen, but these are not of general enough occurrence to allow ^{14}N NMR to be classed as a high-resolution method, at least in comparison with the NMR of spin 1/2 nuclei.

The ^{15}N nucleus offers a different problem. It does have spin 1/2, so that there is no quadrupole-induced relaxation, but it has a small magnetic moment (1/10 that of the proton), and a low natural abundance of 0.365%. Thus, at the natural-abundance level at constant field, the sensitivity of detection of ^{15}N relative to ^{1}H is down by a factor of 3×10^{-6}. Furthermore, because the gain in signal-to-noise ratio from time-averaging of signals improves only as the square root of the time of observation, then, on the critical time basis, ^{15}N is worse than ^{1}H NMR by a factor of about 10^{-11}! Isotopic enrichment of ^{15}N to the 90% level improves this by a factor of 10^{5}, and the first important surveys of the possibilities of ^{15}N for structural analysis used enriched materials.[11] Nonetheless, although ^{15}N-enriched samples give excellent spectra,[12] the history of ^{13}C NMR shows that rapid and general progress simply does not occur until spectra can be taken routinely at the natural-abundance level. One of the first natural-abundance spectra of ^{15}N was of 85% hydrazine taken with time averaging using a 10-mm sample tube without proton decoupling and continuous-wave excita-

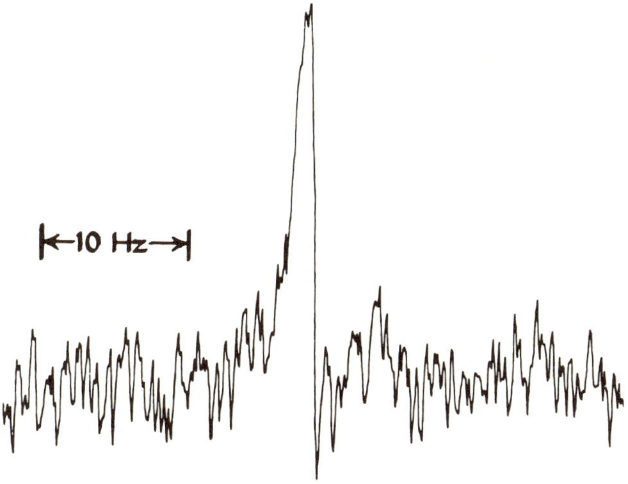

FIG. 1. TIME-AVERAGED, continuous-wave, natural-abundance ^{15}N spectrum of 85% diazane at 6.07 MHz, contained in a 10-mm sample tube, taken with 303 scans at 50 sec per scan.[13]

tion at 6.07 MHz (figure 1).[13] This is essentially an optimum sample through being about 40 M in nitrogen, but it still required a few hours to give an acceptable signal-to-noise ratio.

Further progress became possible through proton decoupling for samples with favorable Nuclear Overhauser Enhancement (NOE), which usually means having nitrogens with attached hydrogens and reasonably short T_1 relaxation times. The NOE for nitrogen is different from that of carbon in that it results in an emission rather than an absorption signal.[14] Thus, depending on the efficiency of the NOE, a ^{15}N resonance may be upright, reduced to insignificance by having absorption just cancel emission, or strongly emissive (maximum = -2.93 times the signal strength of the unsplit absorption signal). The T_1 relaxation times of ^{15}N are highly variable, being greatly influenced by structure and by traces of paramagnetic impurities (especially for water solutions of compounds with nitrogens that carry unshared electron pairs).

Many T_1's for ^{15}N, even those with directly bonded hydrogens, such as simple amines, are in the range from 30-100 seconds, while those of tertiary amines may be on the order of 10^3 seconds. Such relaxation times make taking spectra by the FT technique rather inefficient. The use of paramagnetic relaxing agents is often recommended to shorten the T_1 values, but this is not always helpful because 1) any favorable NOE is cancelled, 2) the linewidths are often increased so much that the signal/noise ratios are not much improved, and 3) there can be significant changes in chemical shift as the result of contact- or pseudocontact-shift contributions resulting from interaction of the paramagnetic species with the nitrogens whose relaxation times are desired to be shortened. Fortunately, the efficiency of dipolar relaxation of ^{15}N by directly attached or neighboring protons increases substantially on slowing molecular reorientations so that as the correlation time, τ_c, becomes greater (at least to about 10^{-8} seconds), the efficiency of taking FT spectra greatly improves. This is especially important for large, biochemically interesting molecules.[15]

The technology that moved natural-abundance ^{15}N NMR into the arena of more routine instrumental techniques was development of relatively large-bore superconducting magnets with a high degree of homogeneity, ~5 parts in 10^9 over a substantial volume, thus permitting more nuclei to be placed within the active volume of the probe—in essence, the "milk-bottle" approach, the feasibility of which was demonstrated at lower magnetic fields for ^{13}C by Allerhand and coworkers.[16] The one serious difficulty with this approach is the very large sample sizes required. The WH-180 spectrometer developed for our research at the California Institute of Technology by Bruker Magnetics works very well with 25-mm sample tubes and, with that size of tube, requires a minimum of 18 ml of sample. Use of 30-mm tubes, with about 40 ml of sample, gave no advantage for routine spectra because of reduced field homogeneity.

One might well argue that there is nothing "routine" about a technique requiring 20 ml of sample, but this is surely a value judgment because it depends so much on the nature of the material, especially on the ^{15}N relaxation times, and also on the length of time the spectroscopist is willing to devote to taking spectra. Thus, as regards time, if the ratio of signal-to-noise for a given set of resonances is required to be improved by a factor of 5 after 2 hours of taking a spectrum, the sample has to be an important one to devote 50 hours to achieve this objective. As mentioned earlier, the relaxation-time problem can be acute for small molecules and even neat *N*-methylazacyclopentane (~10 M) gives only a very weak signal when pulsed with 1-2 minute repetition rates.[17] Contrariwise, 0.08 M cyanocobalamine, a very much larger molecule whose amide nitrogen nuclei relax with greater efficiency, because of a greater τ_c, can be pulsed with a repetition rate of 0.82 second, and gives an excellent spectrum of these nitrogens after 18 hours (figure 2). A still greater effect is observed for the enzyme lysozyme, which at 0.009 M with a 0.82 sec repetition rate was found to give a useful spectrum in 21 hours. For this substance at the given concentration, the relaxation times may be close to the optimum values for rapid pulsing, without serious line broadening. Nonetheless these same relaxation times may be sufficiently short as to quench the NOE's of some of the peptide nitrogens (note the upward resonances of figure 3).

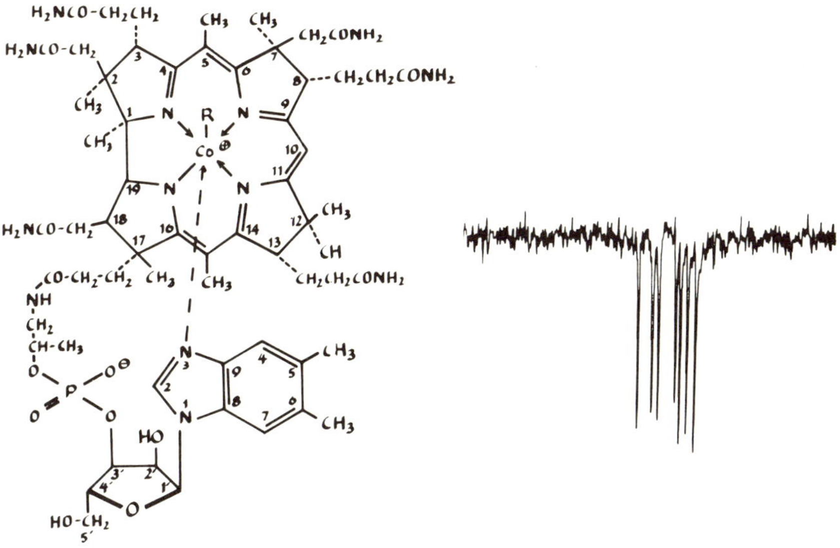

FIG. 2. PULSED FOURIER-TRANSFORM NMR spectrum of cyanocobalamine (vitamin B_{12}) as an 0.08 M solution in 1:1 ethanol-water at 18.25 MHz. The spectrum was taken with 3 mW of proton decoupling, 78,811 pulses (90° flip angle), and a repetition rate of 0.82 sec. The only peaks observed are those that can be attributed to the seven amide ($-CONH_2$) groups.[15]

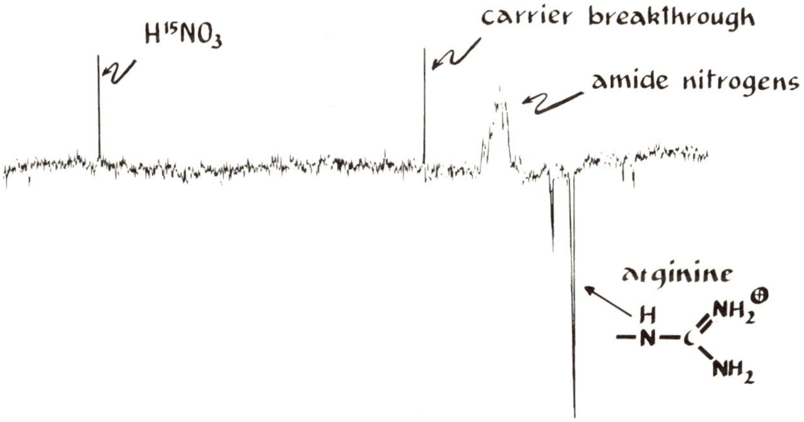

FIG. 3. NATURAL-ABUNDANCE ^{15}N NMR spectrum of 0.009 M egg-white lysozyme in water at pH 3.9. A total of 93,287 transients was taken, with the other spectral parameters being the same as in figure 2.[15]

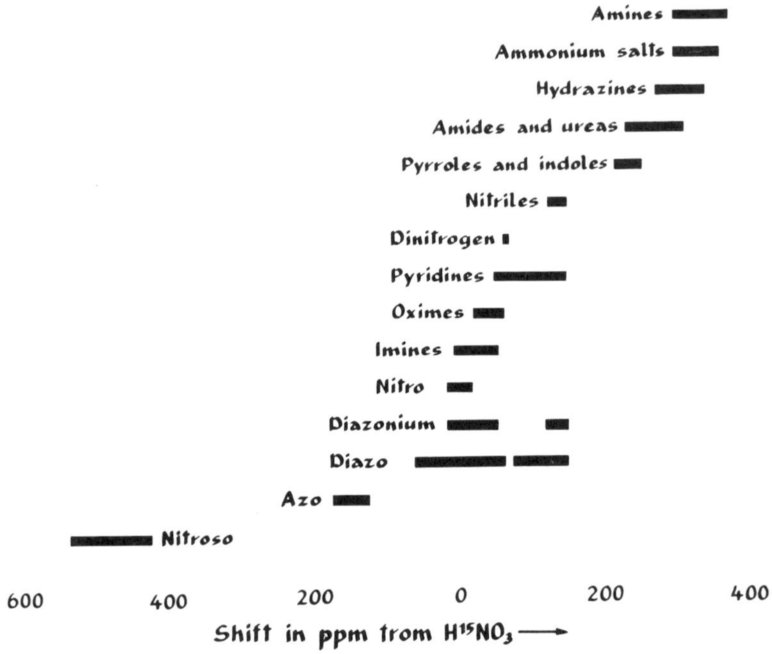

FIG. 4. SOME RANGES OBSERVED for nitrogen chemical shifts as a function of structure.

When the principal obstacle to widespread use of ^{15}N was overcome and literally thousands of substances could be studied, an essentially explosive development period ensued to which many investigators have contributed—most notably Randall (Queen Mary College), Lichter (Hunter College), Levy (Florida State University), and their coworkers. No attempt will be made here to provide a comprehensive review of the current status of ^{15}N NMR. Instead, only a few topics will be covered, largely drawn from research on steric, hydrogen-bonding, protonation, and complexation effects on ^{15}N chemical shifts and the relation between ^{15}N and ^{13}C shifts in analogously constituted compounds.

Structural elucidation assisted by ^{15}N NMR depends very largely on chemical-shift differences. Most spin-spin couplings involving ^{15}N, except for those arising from directly bonded (and not rapidly exchanging) protons, are only a few Hz because of the small magnetic moment of ^{15}N. Thus they usually have limited utility for structural analysis. Chemical shifts of nitrogen span a very considerable range (figure 4), however, as is typical of magnetic nuclei of atoms such as C, F, P, and so on (but not H or Li) that use *p* orbitals in bond formation. The importance of *p* orbitals is that, under the influence of a magnetic field, electrons in these orbitals undergo a forced circulation that can generate a very sizable paramagnetic

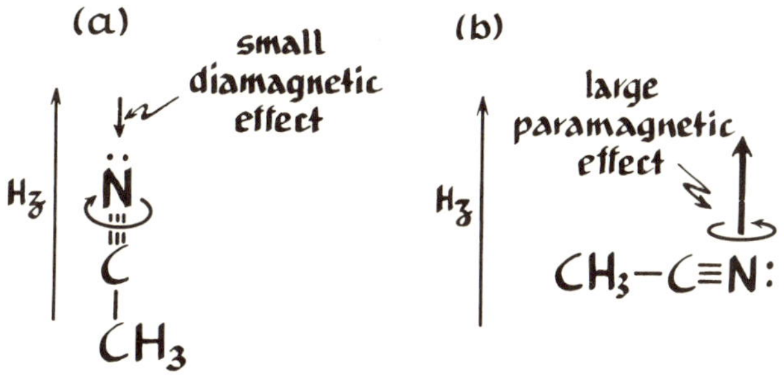

FIG. 5. DIAMAGNETIC AND PARAMAGNETIC ELECTRON CIRCULATIONS postulated for ethanenitrile as a function of orientation in a magnetic field H_z.

effect ("second-order paramagnetic effect") directly at the nucleus undergoing absorption. Because the bonding orbitals normally have a full complement of electrons, circulation of electrons in these filled orbitals produces diamagnetic shielding. To account for the paramagnetic circulation, it is assumed that excited-state wave functions contribute sufficiently to the ground-state electronic configuration to give the observed paramagnetic influences. In this context, the idea that an "electron-promotion" energy should be important to the shifts is a very natural one.

It should be recognized that these paramagnetic circulations do not correspond to the diamagnetic circulations that are used to account for the shifts of the protons of benzene or ethyne. Thus, for a triple bond such as in ethanenitrile, when the molecule is *parallel* to the magnetic field, diamagnetic circulation around the bond produces a diamagnetic effect along the bond axis (figure 5a). However, as has been shown by Pople,[18] this diamagnetic effect is too small to account for the shift of ethyne protons and must also be too small to have much influence on the shift of ^{15}N in ethanenitrile. The second-order paramagnetic effect operates most effectively when the long axis of the ethanenitrile molecule is *perpendicular* to the magnetic field, and at least some (and probably a lot) of the circulation appears to involve one of the electrons of the unshared electron pair on nitrogen and a promotion energy that reasonably corresponds to the $n \rightarrow \pi^*$ excitation energy (figure 5b). The second-order paramagnetic effect is in agreement with the fact that the ^{15}N shift of $CH_3C{\equiv}N$: is about 235 ppm *downfield* from the shift of $(CH_3)_3N$: for which there is no low-lying π^* excited state. Furthermore, when $CH_3C{\equiv}N$ is converted by strong acid to the corresponding conjugate acid, $CH_3C{\equiv}\overset{\oplus}{N}H$, in which the nitrogen resonance is expected by simple theory to be shifted *downfield* because of its positive charge, the opposite occurs and the nitrogen resonance shifts *upfield* by more than 100 ppm.[19] This shift change is very reasonable on the basis of the second-order paramagnetic effect because the $n \rightarrow \pi^*$ transition in $CH_3C{\equiv}N$: becomes a $\sigma \rightarrow \pi^*$ transition in $CH_3C{\equiv}\overset{\oplus}{N}-H$, which should have a much greater energy, thus greatly reducing mixing into the ground state of excited-state electronic configurations that correspond to $n \rightarrow \pi^*$ (or similar) transitions. Further confirmation of the general correctness of these formulations comes from the fact that protonation of $(CH_3)_3N$: results in a *downfield* shift of 13 ppm.[20]

The message of the discussion above is that a nitrogen with an unshared pair that is part of a π-electron system can be expected to have a substantial downfield shift compared to that of a saturated nitrogen. Also, if the unshared pair is protonated or otherwise perturbed so as to increase the promotion energy, then there should be a sizable upfield shift. We will see how this works out in practice.

Before proceeding further, I should mention the shift scales and reference compounds for ^{15}N chemical shifts, because there is no uniformity about this at present. One problem is that, as yet, no one has come up with a suitably inert, easily labeled, widely soluble nitrogen analog of tetramethylsilane to use as an internal shift standard. Consequently, external standards are almost universally used, and in our laboratory we have found it very convenient to use a 1 M solution of ^{15}N-labeled nitric acid in deuterium oxide as an external standard, contained in a 5-mm tube centered in our 20-mm sample tubes. This mixture has the advantage of having deuterium nuclei available to provide a field-frequency lock signal. It is disadvantageous because the position of the reference should be corrected for temperature. Nitromethane labeled with ^{15}N and substituted with deuterium (CD_3NO_2) is a less temperature-sensitive reference, but one that is more difficult and expensive to prepare. The resonance of our nitric-acid standard is 6.2 ppm upfield from neat nitromethane.

Another unresolved problem with using the resonances of either nitromethane or nitric acid as a reference is that the vast majority of ^{15}N shifts are then upfield. Initially, NMR spectroscopists liked downfield standards, the epitome of which would be the bare gaseous nuclei. For proton spectra, water and benzene were once widely used; for ^{13}C spectra, carbon disulfide was the reference of choice, and so on. However, the advent of tetramethysilane as a reference gave a convenient upfield standard for both proton and carbon spectra, and finally some misguided committees decided that *all* NMR shifts should be reported in exactly the same way. Adherence to this dictum means either that all shifts upfield of a standard of reference must be given *negative signs* (a potent device for promoting misunderstanding and mistakes), or else that everything must be referenced to an upfield standard, however inconvenient on other grounds. The present situation is far from desirable, but the best solution is more likely to be achieved by further research rather than by rules set up by committees or journal editors unrepresentative of the workers in the field.

We will be mostly interested in this paper in ^{15}N shift *differences*, but where shifts are given numerical values, these will be in ppm *upfield* from our external nitric-acid standard. To convert to ppm *upfield* from nitromethane, add 6.2 ppm. For ppm *downfield* from 2 M urea in water, 2 M tetramethylammonium chloride in water, and 2 M ammonium nitrate (ammonium resonance) in water, use the relationship $-(\delta_{H^{15}NO_3} - \delta_{STND})$, where δ_{STND} for each of the substances above is 298.7, 332.8, and 355.0 ppm, respectively.[21]

In 1967, Spielvogel and Purser[22] made the very fruitful discovery that there is a correlation between ^{11}B NMR shifts and ^{13}C shifts in reasonably corresponding compounds. Applied to ^{15}N shifts, what this means is that one might expect the ^{13}C shift of C1 of butane to correlate with the nitrogen shift of propanamine, the ^{13}C shift of C2 of 2-methylbutane with the nitrogen shift of *N*, *N*-dimethylethanamine and so on. This possibility has been carefully investigated and found to have great merit for a variety of primary and secondary amines (figure 6).[23] The dependence is linear to a very high degree over about 90 ppm, but, for reasons not yet fully understood, primary and secondary amines are found to fall along lines separated by about 10 ppm, at ^{15}N shifts near 330 ppm, from nitric acid with somewhat different slopes. The slopes of the correlation lines show the ^{15}N shifts are about twice as sensitive to variations in structure as are ^{13}C shifts.

An important point about the amine shifts correlated in figure 6 is that, in all save one of the amines, the nitrogens are not part of a π-electron system. The exception, benzenamine (compound **1**), is expected to be different because of possible conjugation of the unshared pair of electrons on nitrogen with the benzene ring. This sort of conjugation is not expected for the methyl carbon of methylbenzene, and thus it is not surprising that

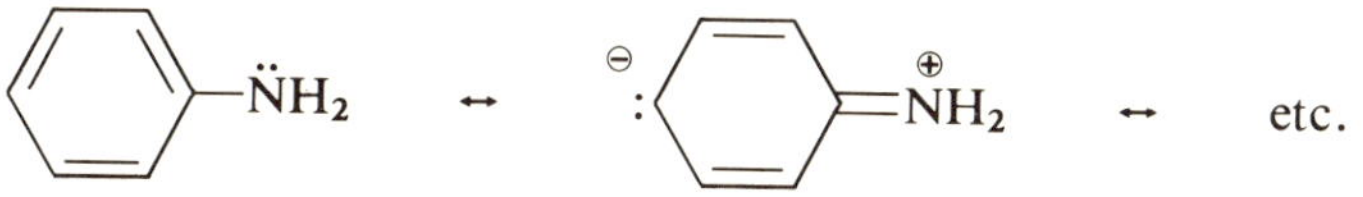

1

the ^{15}N shift of benzenamine is about 18 ppm farther downfield than expected from the correlation line. We can also understand why conversion of compound **1** to benzenammonium ion causes an upfield shift of its ^{15}N resonance.[11]

At the time the data for figure 6 were being collected, it was not possible to obtain useful ^{15}N spectra of tertiary amines, and it could not be determined whether such amines would give a third correlation line in a plot such as figure 6. Later research with a substantial number of substituted

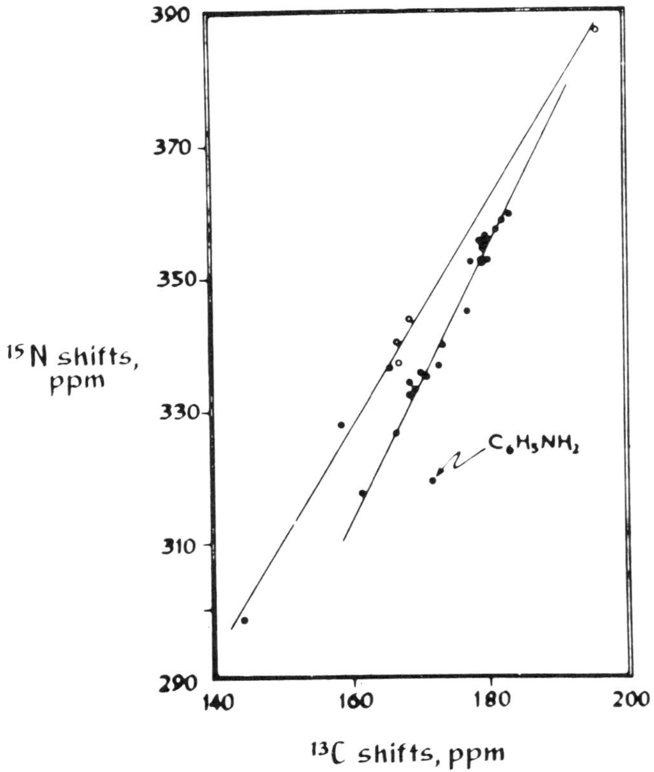

FIG. 6. CORRELATIONS OF ^{15}N CHEMICAL SHIFTS of amines with ^{13}C chemical shifts in analogously constituted compounds. The lower line is for primary amines (RNH_2) and the upper line for secondary amines (RNHR′, where R and R′ may be the same or different). The point marked $C_6H_5NH_2$ is that for benzenamine.[23]

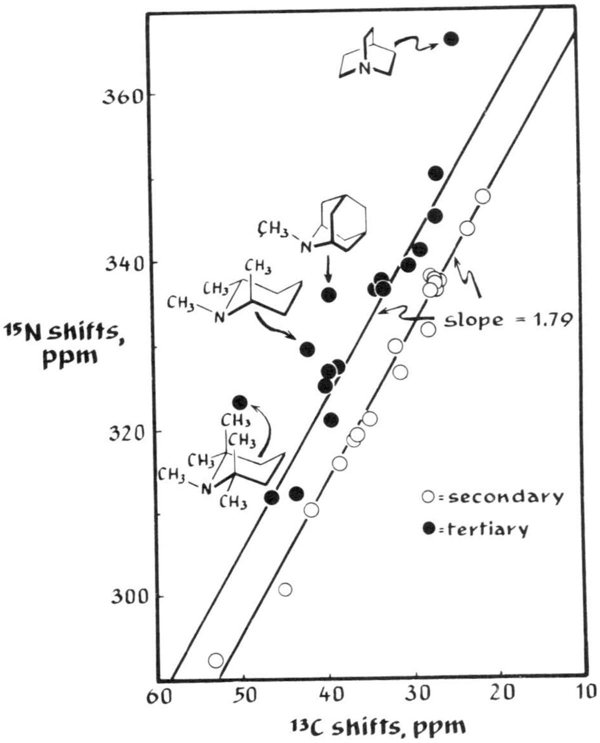

FIG. 7. CORRELATION OF ^{15}N AND ^{13}C SHIFTS for various secondary and tertiary azacyclohexanes.[21] For all of the compounds shown, the nitrogen is included in at least one cyclohexane ring.

azacyclohexanes and aliphatic tertiary amines has clearly shown that tertiary amines give more complex behavior.

In the first place, it will be seen from figure 7 that, even excluding four special cases to be discussed below, the shifts of the tertiary azacyclohexanes show rather more scatter than the secondary azacyclohexanes when one tries to correlate their ^{15}N shifts with the ^{13}C shifts of analogously substituted hydrocarbons.[24] There are a number of possible reasons for this, but certainly an important one is the fact that an *N*-substituted azacyclohexane usually has more conformational freedom than the corresponding hydrocarbon. For example, while *trans-trans*-1, 2, 3-trimethylcyclohexane (compound **2**) has two possible conformations, with **2a** (three equatorial methyl groups) being expected to be favored over **2b** (three axial methyl

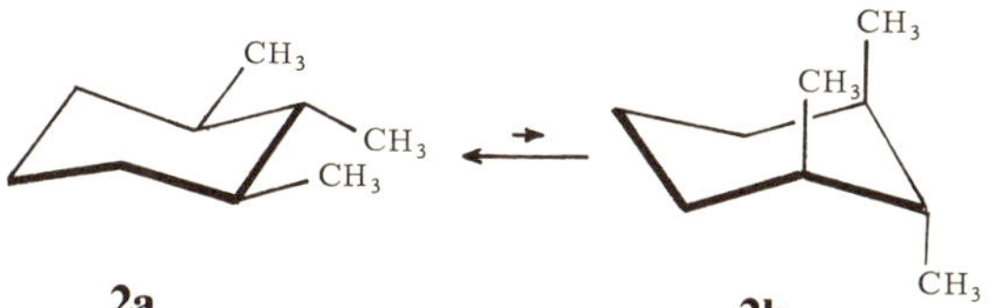

2a **2b**

groups), *N*-methyl-*cis*-2, 6-dimethylazacyclohexane (compound **3**) has four possible equilibrating conformations. Of these, conformer **3a** is expected to be more favorable, but conformer **3b** is very likely to make up a significant fraction of the equilibrium population and, if this is so, the ^{15}N shift will surely be perturbed toward higher fields, away from the correlation line (as is observed), as compared to the ^{13}C shift of compound **2**.

3a **3b**

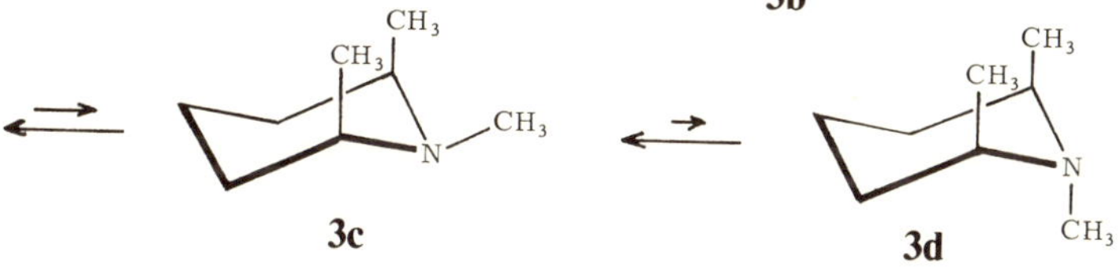

3c **3d**

It is unlikely that conformational mobility is the whole story of deviations of the ^{15}N shifts from the correlation line with ^{13}C shifts. The ^{15}N resonance of the conformationally rather rigid 1-azabicyclo[2.2.2]octane (quinuclidine, compound **4**) is 15 ppm away from what would be expected from the C1 shift of bicyclo[2.2.2]octane (compound **5**), while the ^{15}N shift of the at-least-as-rigid *N*-methyl-2-azabicyclo[3.3.1.1^{3,7}]decane (2-azaadamantane, compound **6**) is 13 ppm from predictions based on C2 of 2-

methyltricyclo[3.3.3.1.1^{3,7}]decane (adamantane, compound **7**).

4

5

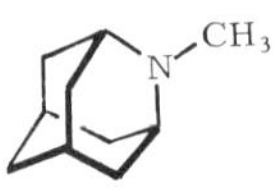

6

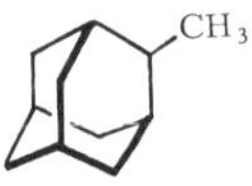

7

These deviations suggest that a substantial upfield shift is associated with an *antiperiplanar* orientation of the nitrogen electron-pair orbital with respect to adjacent C-C (but not C-H) bonding orbitals, as in partial structure **8**.[21]

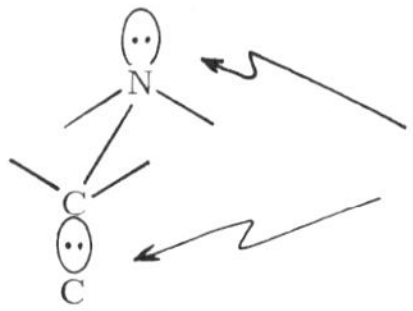

8

antiperiplanar orientation of the orbital for the electron-pair of nitrogen and an adjacent C-C bonding orbital.

It will be seen that structure **4** has three such orientations, while structure **5** has two. Two facts support the importance of these orientations. One is that a variety of azacyclohexanes, for example 3(e)- and 9(a)-methyl-1-azabicyclo[4.4.0]decane (3(e)- and 3(a)-methylquinolizidines, compounds **9** and **10**), which do not have bond arrangements such as structure **8**, have ^{15}N shifts that correlate with the ^{13}C shifts of the corresponding hydrocarbons.

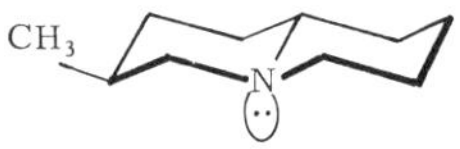

9

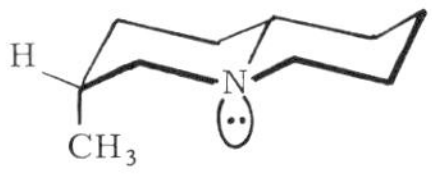

10

The other fact is that compound **4**, compound **6**, and *N*-methyl-*trans*-2, 6-dimethylazacyclohexane (compound **11**), and *N*-methyl-2, 2, 6, 6-tetramethylazacyclohexane (compound **12**) (all with one or more anti-

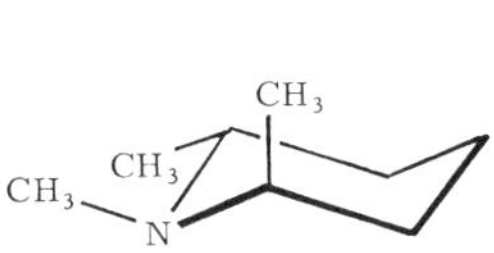

11

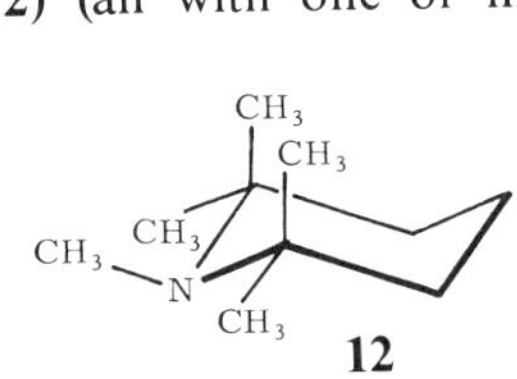

12

periplanar arrangements corresponding to structure **8**) have much larger hydrogen-bonding and protonation shifts of their nitrogens than do the azacyclohexanes, which do not have favorable conformations for arrangements such as structure **8**. This means that, as the electron pair on nitrogen gets tied up in bond formation, the electronic interaction corresponding to **8** gets to be less important, and the anomalous ^{15}N shifts tend to disappear.

The fact that secondary azacyclohexanes such as *trans*-2, 6-dimethyl- and 2, 2, 6, 6-tetramethylazacyclohexane do not show the *antiperiplanar* effect is the result either of some special influence of the hydrogen bond to nitrogen, or else of the very simple idea that, in these substances, the N-H hydrogen in the equation below is largely axial, a situation that would make the unshared-pair orbital *uniformly antiperiplanar* to the ring C-C orbitals. This possibility seems unlikely from what is known of the axial-equatorial equilibrium of the N-H proton of azacyclohexanes.[25]

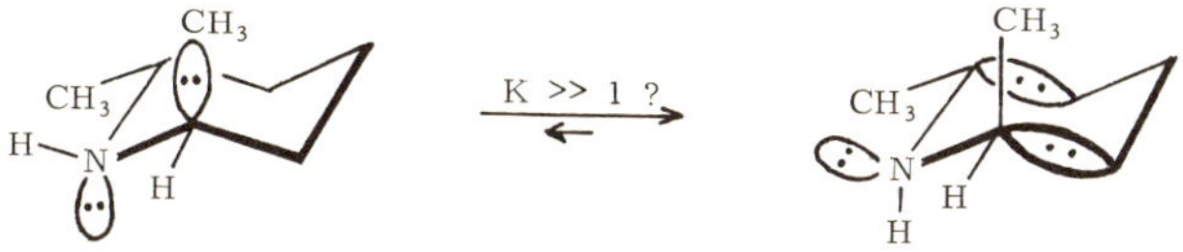

(1)

antiperiplanar arrangement of N and CH_3 orbitals

antiperiplanar arrangement of nitrogen and ring C-C orbitals

One possible difference in behavior between tertiary and secondary azacyclohexanes is that molecules of the latter can aggregate by hydrogen bonding to each other, >N—H · · · · :N(H)< . This does not suffice to account for the substantial differences in general behavior between these classes of amines, because changes in solvent and concentrations that would be expected to change the hydrogen bonding generally produce effects on the ^{15}N shifts of the secondary azacyclohexanes only on the order of $\pm 1 - 2$ ppm.

Aliphatic tertiary amines present a very different picture from the tertiary azacyclohexanes,[24] as can be seen from figure 8. The scatter of the points from the correlation line is relatively large, and also the slope of 1.39 of the correlation line is very much less than the value of 1.79 for tertiary azacyclohexanes shown in figure 7. The reason for this is not clear, but could well be related to the fact that the rings of azacyclohexanes restrict their tertiary brethren to conformations similar to those of the corresponding hydrocarbons, while aliphatic tertiary amines may not be so constrained.

There are further points of interest about the amine shifts. First, the ^{15}N-^{13}C correlation lines for the conjugate acids of primary and secondary amines have smaller slopes than those of the primary and secondary amines themselves (see figures 9 and 10). This would appear to indicate that the more closely the amine resembles the hydrocarbon in bond structure, the closer the slopes of the correlation line between ^{13}C and ^{15}N shifts approach unity. This suggests that much of the larger sensitivity of ^{15}N shifts to structural changes when compared with ^{13}C shifts is a property of the unshared pair on nitrogen, and, possibly, that the polarizability of this electron pair is

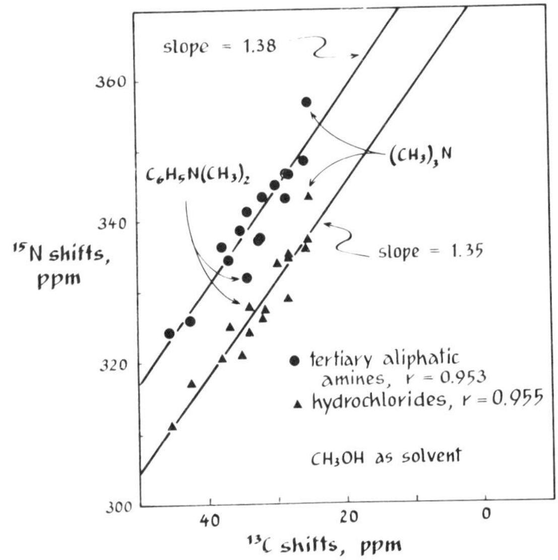

FIG. 8. CORRELATION OF ^{15}N AND ^{13}C shifts of tertiary aliphatic amines and their hydrogen-chloride salts in CH_3OH as solvent.[24]

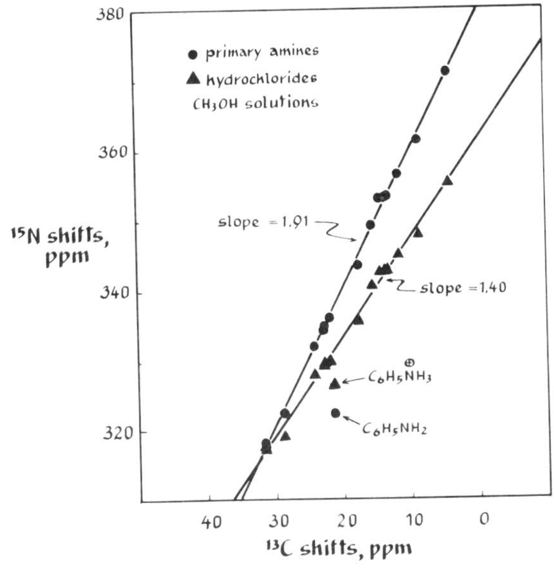

FIG. 9. CORRELATION OF ^{15}N AND ^{13}C SHIFTS of primary aliphatic amines and their hydrogen-chloride salts in CH_3OH solutions.[24]

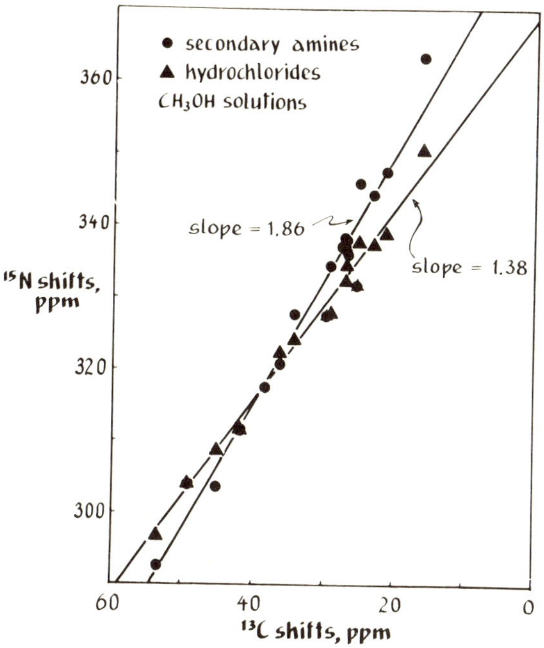

FIG. 10. CORRELATION OF ^{15}N AND ^{13}C SHIFTS of secondary aliphatic amines and their hydrogen-chloride salts in CH_3OH as solvent.[24]

the important variable. It is interesting in this connection that the slopes of the correlation lines of the amine salts we have studied (including tertiary amines) have slopes within the narrow range of 1.35-1.40, although somewhat different intercepts.[24]

It will be seen from figures 9 and 10 that the ^{15}N-^{13}C correlation lines for primary and secondary amines (but not tertiary amines, see figure 8) and their salts cross over. Because the correlations are linear, this leads to the curious fact that the protonation shifts are linear with the ^{15}N shift of the amine itself and, further, if the ^{15}N shift of the amine is far downfield, protonation will result in an *upfield* shift of the nitrogen. The trends are shown in figures 11 and 12, and it turns out that all of the primary and secondary amines in these correlations that give small or *upfield* protonation shifts have branching at the carbons (the α carbons) to which the nitrogens are attached, and the largest *downfield* protonation shifts of the compounds correlated in figures 11 and 12 are CH_3NH_2 and $(CH_3)_2NH$ with no branching. It will be seen that the ^{15}N shifts of these amines with branching at the α carbons tend to behave something like that of benzenamine (compound **1**), in being downfield of the simpler amines and tending to give small or upfield protonation shifts.

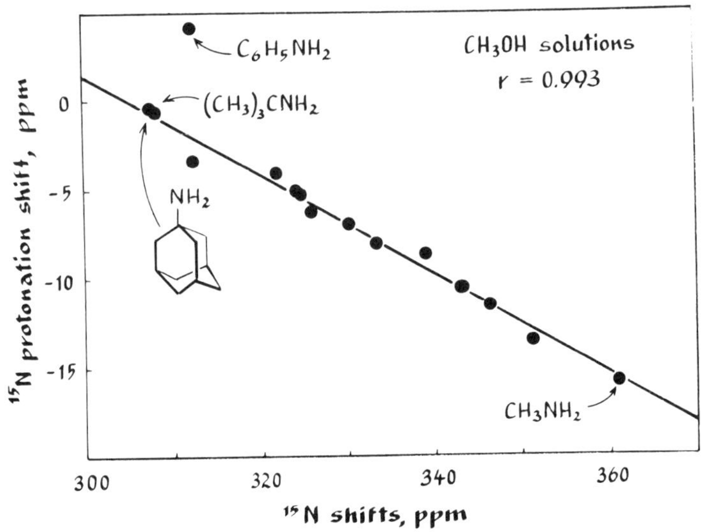

FIG. 11. CORRELATION OF ^{15}N SHIFTS of primary amines with ^{15}N protonation shifts of the same amines, CH_3OH solutions.[24]

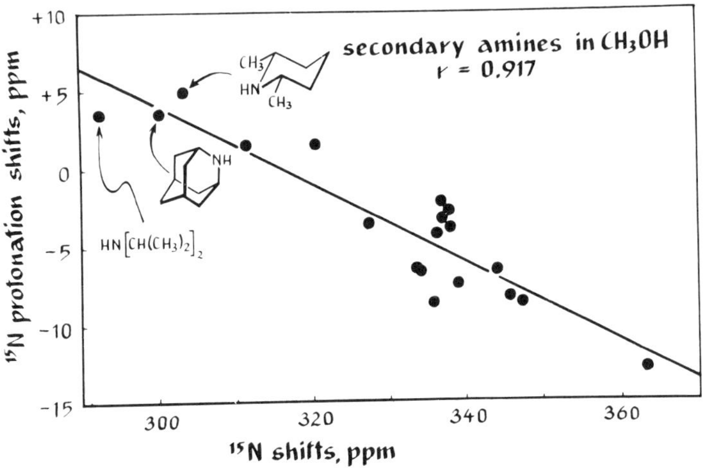

FIG. 12. CORRELATION OF ^{15}N SHIFTS of secondary amines with ^{15}N protonation shifts of the same amines, CH_3OH solutions.[24]

The difference in behavior of tertiary amines, for which we postulate important effects arising from the *antiperiplanar* arrangements of groups (**4**, **6**, **11**, and **12**), is strikingly evident in their protonation shifts, which are large (14 to 24 ppm) but in the *downfield* direction, even though **6**, **11**, and **12** possess the very branching at the carbons to which the nitrogens are attached, which is seen in figures 11 and 12 to lead to protonation shifts in the *upfield* direction for primary and secondary amines.[24]

Influences of solvents on the ^{15}N shifts of primary, secondary, and tertiary amines, as well as those of solvents, concentrations, and counterions on amine salts have been investigated in a systematic way.[24] Some large effects have been noted. Thus, for the conjugate acid of *cis*-3, 5-dimethylazacyclohexane (compound **13**), with the counterion $X^{\ominus}=Cl^{\ominus}$, there is a downfield shift of 2.2 ppm when the concentration in $CHCl_3$ is increased from 7.7 to 10.8 mole %.

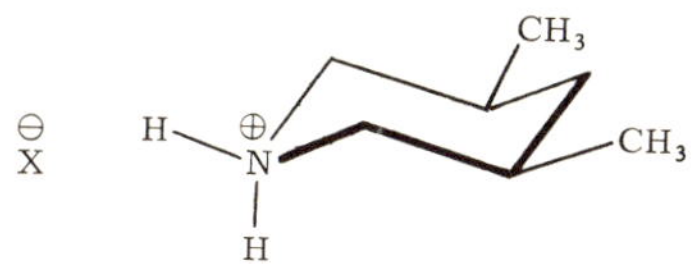

13

Change of $X^{\ominus}$ from $Cl^{\ominus}$ to $I^{\ominus}$ in $CHCl_3$ results in a downfield shift of 3.8 ppm, and from $Cl^{\ominus}$ to $BF_4^{\ominus}$ in an upfield shift of 7.9 ppm, and thus a shift difference from $X=I^{\ominus}$ to $BF_4^{\ominus}$ of 11.7 ppm! The influence of $X^{\ominus}$ in CH_3OH as solvent is much smaller, and with $X^{\ominus}=Cl^{\ominus}$, $I^{\ominus}$, and $BF_4^{\ominus}$, the shift differences are reduced to 2.9 ppm, with $I^{\ominus}$ and $BF_4^{\ominus}$ only 2.3 ppm apart.[24] With azoniabenzene (pyridinium) salts, it has been found that the solvent and counterion effects on the *N*-alkyl salts are much smaller than on the *N*-hydrogen salts.[26]

Another point of interest, and a highly practical one for structural elucidation, is that ^{15}N shifts of azacyclohexanes exhibit the pattern of shift differences where axial and equatorial substituents are involved, a pattern that is such a useful feature of ^{13}C NMR spectra.[4b,27] The correlation of figure 7 requires that this be so for nitrogens because many of the substances that correlate well have 3-axial methyl substituents (figure 13).[21,28] That similar differences are displayed for an amine group axial or equatorial (upfield by >5 ppm for axial nitrogen) is demonstrated by the *cis*- and *trans*-4-*tert*-butylcyclohexanamines (figure 13) and by data for 2-amino-2-deoxy-D-hexose derivatives.[29]

For virtually all compounds of which one would have sufficient material to take a natural-abundance ^{15}N spectrum, it would be supposed that the structure would have already been established beyond question. This is not always true for substances where we could determine the precise structure by x ray diffraction for the crystalline state, but which might exist in other forms in solution by operation of acid-base or tautomeric equilibria. One such case is sulfaguanidine, for which structure **14** appears in the usual reference works, despite some earlier physical evidence that structure **15** might in fact be correct. The proton-coupled ^{15}N NMR

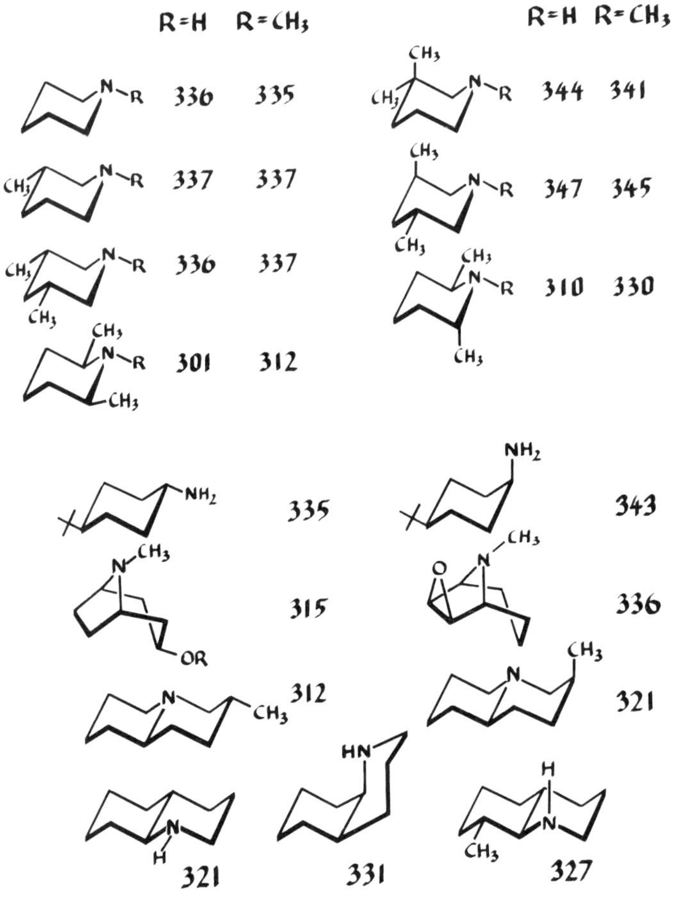

FIG. 13. EFFECT OF AXIAL *VS.* EQUATORIAL ORIENTATIONS on ^{15}N shifts of various azacyclohexanes.[21,24-28] All shifts are in ppm relative to nitric acid as standard.

spectrum (figure 14) is decisive in distinguishing between **14** and **15**.[30] It will be seen that there are two different $-NH_2$ resonances (1:2:1 triplets) with intensity ratios of 1:2 and a strongly downfield unsplit $-N=C\langle$ resonance in agreement with structure **15**. That the $=C(NH_2)_2$ nitrogen resonances appear equivalent with the same chemical shift, despite the possibility of *cis-trans* isomerism, may reflect substantial resonance contributions of the valence-bond structure **16** in a nitrogen-inversion transition state.

Another example is provided by the nine-membered lactam, 1-aza-2-cyclononanone. The studies of Huisgen and coworkers[31] have provided evidence that this substance exists in solution as a mixture of the *cis* and *trans*

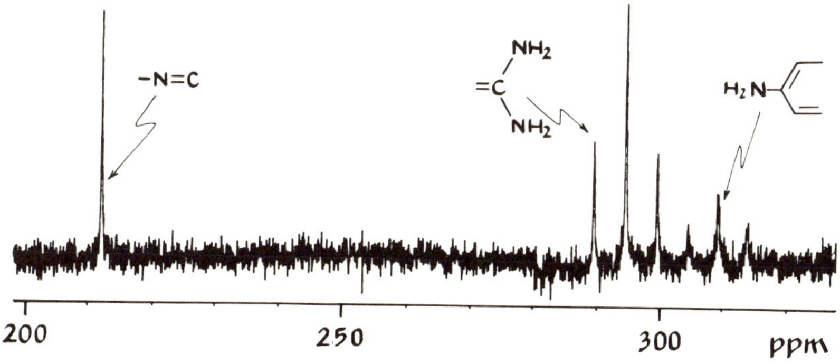

FIG. 14. NATURAL-ABUNDANCE PROTON-COUPLED ^{15}N SHIFTS of sulfaguanidine (structure **15**) in methylsulfinylmethane solution at 18.23 MHz.[30]

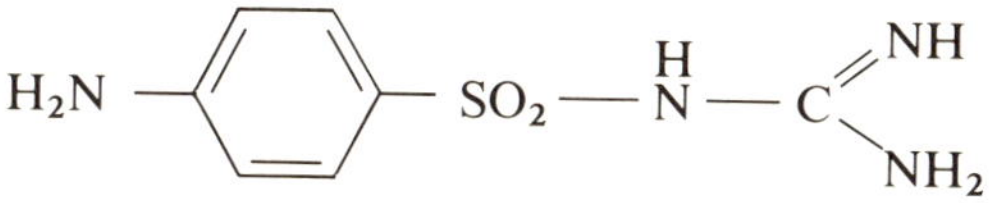

14

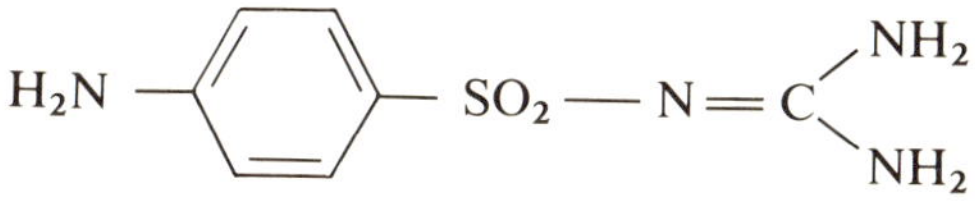

15

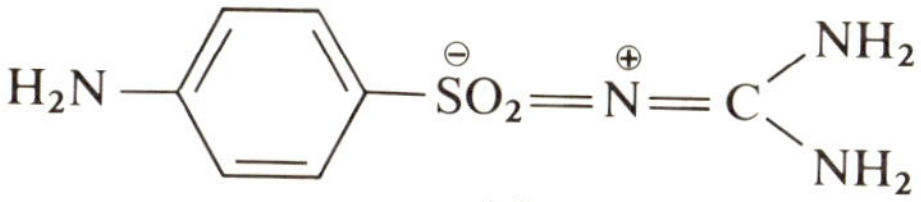

16

isomers **17a** and **17b**, respectively. This despite the fact that the crystalline material is exclusively **17b**.[32] However, the ratio of **17a** to **17b** in solution is a matter of some dispute.[33] When the crystals of **17b** are dissolved in trichloromethane at $-35°$, the ^{13}C spectrum indicated the presence of **17b** in two conformations. When the temperature was raised to $+35°$,

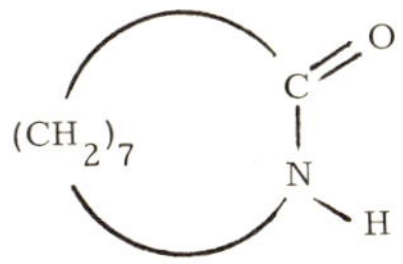

17a

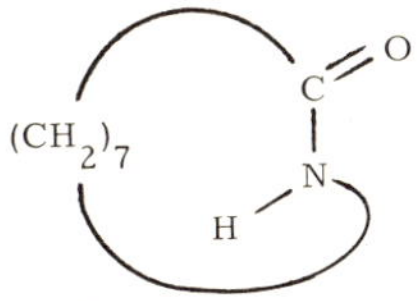

17b

isomerization occurred and ^{15}N resonances were observed separated by 3 ppm for **17a** and **17b** in a ratio of 3.5 to 1. In ethanol, the ratio is close to 1:1.[33]

The relative rates of N-H hydrogen exchange of the *cis-trans* isomer pair, **17a** and **17b**, in acid or basic solution can be very easily determined from the changes in line shape that occur when exchange becomes rapid on the NMR time scale, because the ^{15}NH couplings are washed out (figure 15).[34] This procedure for measuring the N-H exchange rates is very simple and is of special value whenever ^{14}N quadrupole relaxation results in broadening the proton resonance lines to the point where they are not themselves useful in obtaining the exchange rates, as is the case for many amides (most notably the primary ones). The exchange rates of a number of amides of this type have been determined by following changes in line shapes arising from ^{15}NH splitting in water and methylsulfinylmethane solutions.[35] The technique has also been applied to intermolecular proton exchanges of phenyldiazane (structure **18**), where, contrary to a published report, $-NH-$ exchange occurs much more slowly than $-NH_2$ exchanges (see figure 16).[36]

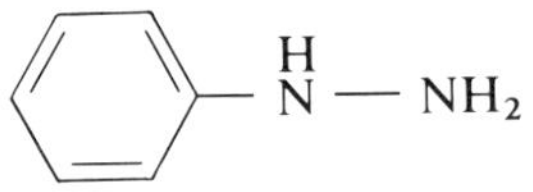

18

A research area of great current interest concerns the interaction between inorganic ions and complexing agents such as the crown ethers and cryptates. We have studied the nitrogen shifts associated with complexation of a number of diamagnetic salts in trichloromethane solution. Some typical results for the *changes* in ^{15}N shift (ppm) on complexation of 1, 10-diaza-4, 7, 13, 16-tetraoxacyclooctadecane (structure **19**), and diaza-4, 7, 13, 16, 21, 24-hexaoxabicyclo[8.8.8]hexacosane (structure **20**) are given in table I (page 168).[37]

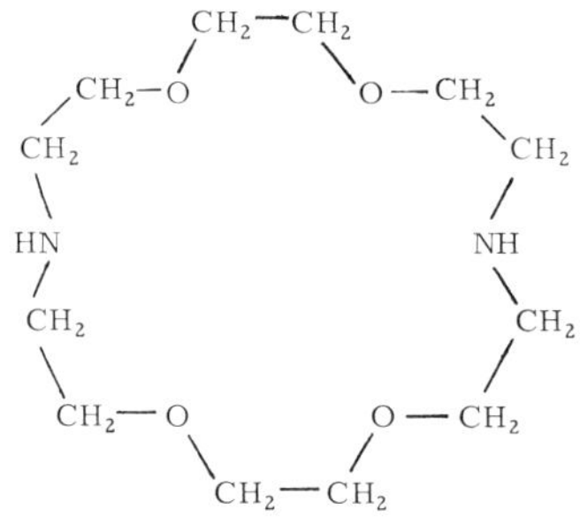

19

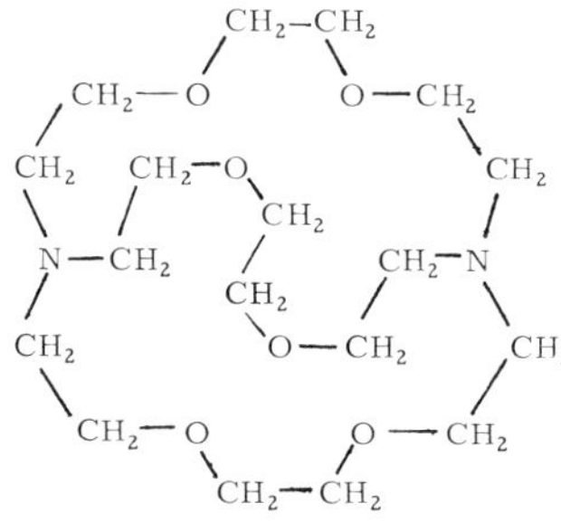

20

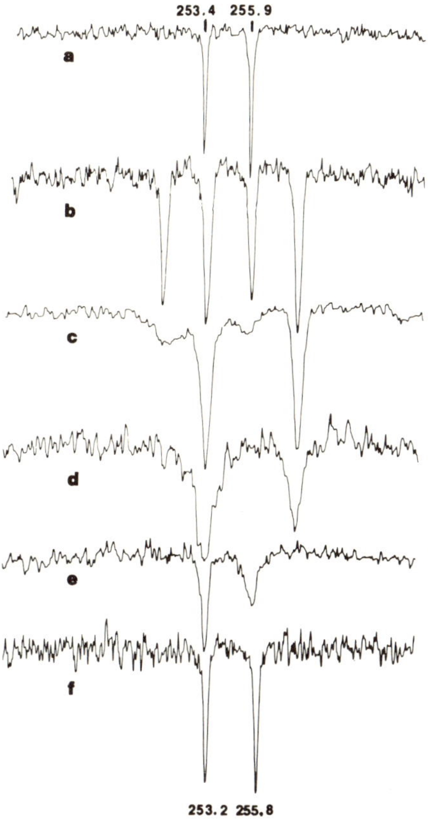

FIG. 15. NATURAL-ABUNDANCE ^{15}N NMR SPECTRA of 2.5 M solution of *cis* and *trans* isomers of 1-aza-2-cyclononanone in methylsulfinylmethane at 25 °C; a) with complete ^{1}H decoupling; b) with gated ^{1}H noise decoupling (to preserve ^{15}N-H spin-spin couplings); c) as in b) but containing 0.008 M NaOH; d) same as b) but containing 0.02 M NaOH; e) as in b) but containing 0.04 M NaOH; f) as in e) but with complete ^{1}H decoupling.[34]

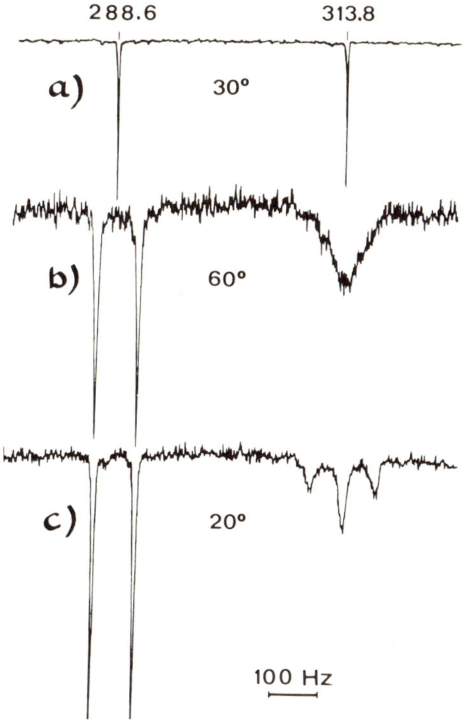

FIG. 16. NATURAL-ABUNDANCE ^{15}N SPECTRA of neat phenyldiazane at 18.23 MHz; a) with ^{1}H noise decoupling; b) and c) with gated ^{1}H noise decoupling at two different temperatures.[36]

These changes in shift are most commonly paramagnetic (negative sign). Diamagnetic shifts are observed with sodium ion and also lithium ion (using other ring sizes of complexing agents corresponding to **19** and **20**). Silver ion, surprisingly, is diamagnetic with **19** but paramagnetic with **20**. One other trend can be inferred by looking at the difference between the shift changes produced by complexation of **19** and **20** with $Na^{\oplus}$, $K^{\oplus}$, $Sr^{2\oplus}$, and $Ba^{2\oplus}$. With these ions, the complexation shift changes for **20** are more diamagnetic by about 3 ppm than those for **19**. Silver ion is much different in having a large paramagnetic complexation shift for the change from **19** to **20**. This difference in behavior probably reflects the propensity of silver ion to form more covalent bonds to amine nitrogen than do alkali or alkaline-earth metal ions. The change from diamagnetic to paramagnetic ^{15}N shifts across the sequence, $Na^{\oplus} \rightarrow Ba^{\oplus}$, has been previously observed for complexes of ethylenediaminetetracetic acid.[38]

TABLE I

^{15}N NMR SHIFT CHANGES ON COMPLEXATION OF THE INDICATED IONS WITH STRUCTURES **19** AND **20**

	$2H^{\oplus}$	$Na^{\oplus}$	$K^{\oplus}$	$Sr^{2\oplus}$	$Ba^{2\oplus}$	$Ag^{\oplus}$	$Tl^{\oplus}$
19	−9.8	2.8	−3.2	−4.0	−11.8	5.0	−17.2
20	−8.7	7.1	−0.1	−2.8	−9.6	−4.8	−16.3

The large paramagnetic shift of multiply-bonded nitrogens carrying an unshared pair of electrons, such as $>C=\ddot{N}-$ or $-C\equiv N:$, and the large upfield shift associated with protonation of the unshared pair were mentioned earlier. This general pattern of shifts has many chemical applications. Of special interest is its use in the study of hydrogen bonding and self association. Hydrogen bonding (and protonation) is known to raise the energy of $n \rightarrow \pi^*$ transitions.[39] We then should expect that hydrogen bonding to the unshared pair on $=\ddot{N}-$ or $\equiv N:$ should result in an upfield (diamagnetic) shift of the ^{15}N resonance. This effect has long been recognized for azabenzene (pyridine, structure **21**) and related azines[40,41] and recently has been studied in some detail.[26]

21

After correction for bulk-susceptibility effects, it turns out that the ^{15}N resonance of **21** in the gas phase is only about 2 ppm downfield of a dilute solution in cyclohexane. There is a substantial upfield shift of 3-4 ppm with the change from cyclohexane to benzene as solvent, or to neat **21**. These shifts suggest charge-transfer or comparable association of the nitrogen lone pair of one molecule of **21** with an aromatic ring of another molecule, which is not unreasonable in view of the fact that **21** boils higher than benzene. The ^{15}N shift of **21** dissolved in methylsulfinylmethane (CH_3SOCH_3), which is a rather polar aprotic solvent and has excellent hydrogen-bond acceptor, but not hydrogen-bond donor, properties, is almost the same as in the neat liquid. This means that the polarity of the solvent *per se* is not a very important characteristic of the solvent in determining the ^{15}N shifts of **21**—a conclusion that is illustrated in more detail by figure 17, where the ^{15}N shifts of **21** are plotted against the so-called reaction-field parameter, which is a function of ϵ, the dielectric constant of the solvent $(\epsilon - 1)/(2\epsilon + 2.5)$. Hydrogen-bond donor solvents cause consistent upfield shifts. Thus, going from cyclohexane to $CHCl_3$ to CH_3OH to H_2O to CF_3CH_2OH gives successive increments of 9.4, 10.6, 2.5, and 10.4 ppm to the ^{15}N shift, respectively—a total shift of 32.9 ppm! Correlation of these shifts with solvent changes on the $n \rightarrow \pi^*$ energies in the optical spectrum of **21** is not easy, because the $n \rightarrow \pi^*$ transition is only visible as a shoulder on the more intense $\pi \rightarrow \pi^*$ absorption, except in nonpolar solvents. However, the work of Kosower and coworkers[42] has shown that solvent-dependent energies (*Z*-values) of the charge-transfer excitation of *N*-ethyl-4-methoxycarbonyl-1-azoniabenzene iodide (structure **22**), gives an excellent linear correlation with the solvent dependence of the $n \rightarrow \pi^*$ excitation energy of cyclohexanone and other solvent-dependent ultraviolet

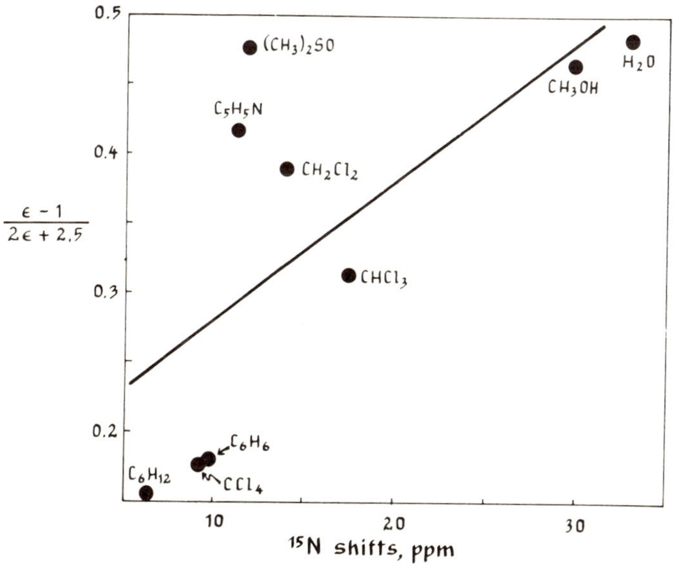

FIG. 17. ATTEMPTED CORRELATION OF ^{15}N SHIFTS of azabenzene in different solvents with a reaction-field parameter that depends on the dielectric constant ϵ of the solvent.[26]

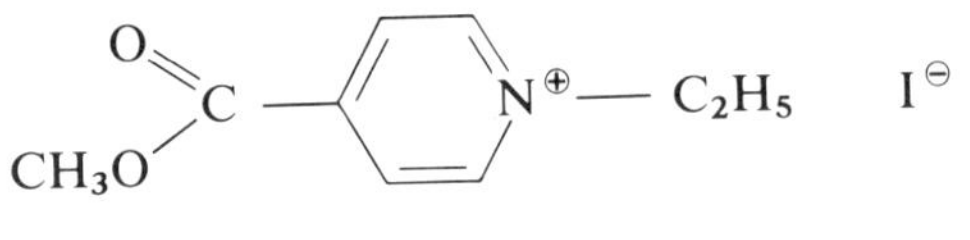

22

absorptions. Figure 18 shows that the trend of the ^{15}N shift of **21** as a function of solvent is reasonably linear with Z ($r = 0.944$).

Perhaps surprisingly, the large protonation shifts of **21** are not very sensitive to solvent. Thus, $CHCl_3$, CH_3SOCH_3, CH_3OH, and H_2O produce an average protonation shift of 99 ± 2 ppm. The point is that one might expect that strong hydrogen bonding would use up a sizable fraction of the total possible upfield shift of the ^{15}N resonance on protonation, so that when one adds a strong acid to **21** in a strongly hydrogen-bonding solvent, the resulting upfield shift increment would be less than in a non-strongly hydrogen-bonding solvent. This does not happen, and strongly hydrogen-bonding solvents shift *both* the resonance of **21** and of its salt upfield. As mentioned earlier, it appears that solvent and counterion effects are smaller for the *N*-alkyl salts of **21** than for the *N*-hydrogen salts.[26]

Alkylideneazanols (oximes, structure **23**) illustrate two important points about the ^{15}N shifts of $\rangle C{=}\ddot{N}\langle$ types of nitrogens.[43]

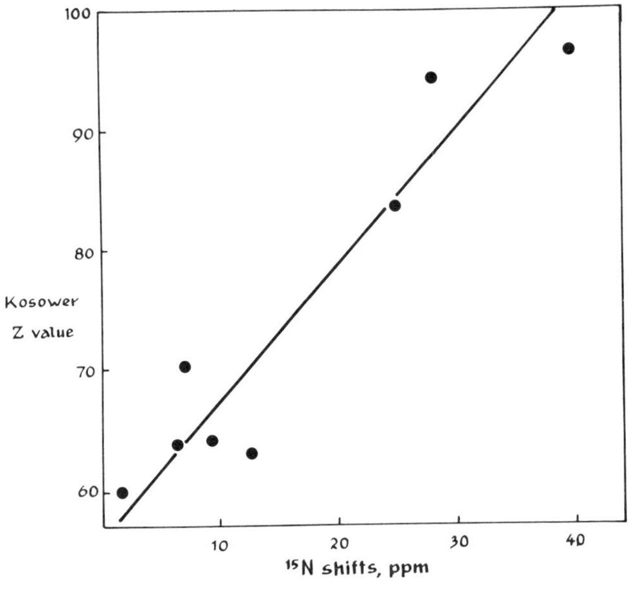

FIG. 18. CORRELATION OF THE ^{15}N SHIFTS of azabenzene in various solvents with the Kosower[42] Z values of the solvents.[26]

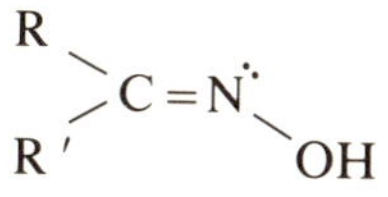

23a

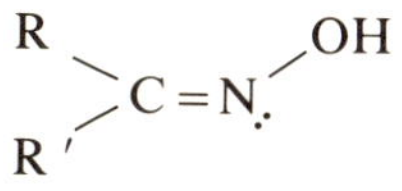

23b

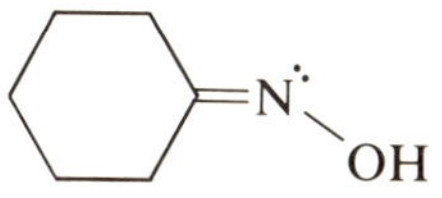

24

First, cyclohexylideneazanol (structure **24**) might seem unusual because the resonance of its nitrogen changes less than 2 ppm in the change of solvent from benzene to $CHCl_3$ to CH_3OH (the corresponding change for azabenzene, structure **21**, is 17.2 ppm). Structure **24** has a fairly strong hydrogen-bonding group in the form of its own OH, however, and can be self-associated as structure **25** or higher aggregates in nonpolar solvents. In CH_3OH, these aggregates may be broken up, with CH_3OH taking over the hydrogen bonding in place of oxime molecules but without much change expected in the ^{15}N shifts. The correctness of this formulation is indicated by two facts. First, when the concentration of **24** is reduced in benzene, the ^{15}N resonance starts to move downfield, as expected for an equilibrium between associated and monomeric oxime. Second, when **24** is dissolved in CH_3SOCH_3, its ^{15}N resonance is 12.2 ppm *downfield* of its value in benzene. What happens is that aggregates such as **25** are broken down because CH_3SOCH_3 is an excellent hydrogen-bond acceptor and soaks up the hydrogen bonds possible with the OH groups of **24** to give **26**, thus freeing the >C=N: unshared pair from hydrogen bonding and causing a downfield shift.

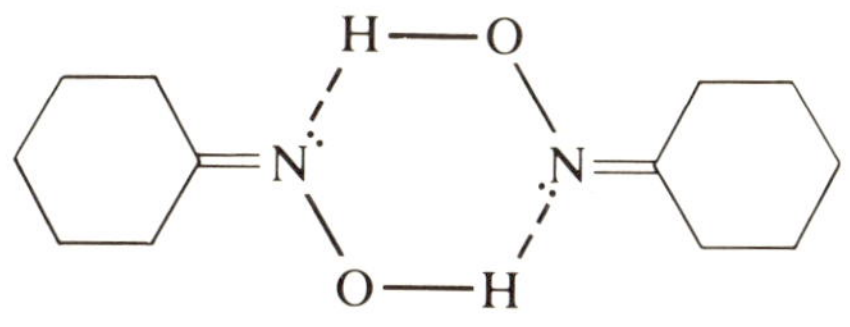

25

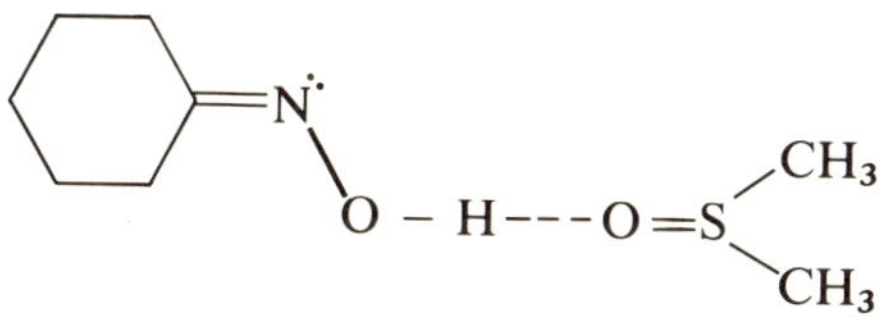

26

The other point of special interest about the aklylidenazanols are the shifts of the *Z* and *E* (*syn* and *anti*) isomers of the type **23a** and **23b**. When R and R′ are different alkyl groups, the difference in nitrogen shifts varies from 2.2 to 5.1 ppm, with larger differences being associated with R and R′

having substantial differences in steric size. Surprisingly, when R = alkyl and R′ = H, the difference in shifts between *E* and *Z* is smaller and sometimes negligible. Thus, with R = $(CH_3)_2CH-$ (**27a**) and R′ = H (**27b**), the shift difference is only 0.1 ppm as compared to 5.1 ppm for R = $(CH_3)_2CH-$ and R′ = CH_3-!

27a **27b**

Apparently, there is a complicated interplay of steric and electrical effects on the ^{15}N shifts of these compounds.

The relation between the ^{13}C shifts of hydrocarbons and ^{15}N shifts of analogously constituted amines has been discussed earlier *ad nauseam* and, in general, the parallelism seems good, with the ^{15}N shifts being about twice as sensitive to substituent effects as the ^{13}C shifts. The possibility of a similar parallelism for the ^{15}N shifts of alkylidenazanols and ^{13}C shifts of alkenes exists. Thus, would the C2 shift of *cis*-2-pentene (structure **28**) correlate with the ^{15}N shift of *Z*-propylidenazanol (structure **29**)? Clearly

28 **29**

there are differences in that **29** has an OH group where **28** has a CH_3, but this should not be a serious obstacle, because it has been shown that an $-OH$ substituted on a hydrocarbon has the same *relative* effect as a CH_3- group on ^{13}C shifts.[44] The really important difference between **28** and **29** is the way that the large downfield nitrogen shift of **29** is associated with a paramagnetic effect arising from mixing into the ground state a contribution of an excited state, or states, corresponding to an $n \rightarrow \pi^*$ optical transition. Obviously, because **28** has no *n* electrons, its ^{13}C shifts cannot be affected in the same way. Figure 19 shows that there is indeed a very good correlation ($r = 0.986$) between the ^{15}N and ^{13}C shifts of pairs of compounds such as **28** and **29** with a slope of 2.18, not much larger than the slopes of amine correlation lines of figures 6-10.[45]

In our present state of knowledge of what determines ^{13}C and ^{15}N shifts, about all, it seems, that can be said about this, is that whatever contribution the $n \rightarrow \pi^*$ excited state makes to the ^{15}N shifts of compounds like **29**, it is not greatly changed by the presence of different substituent groups

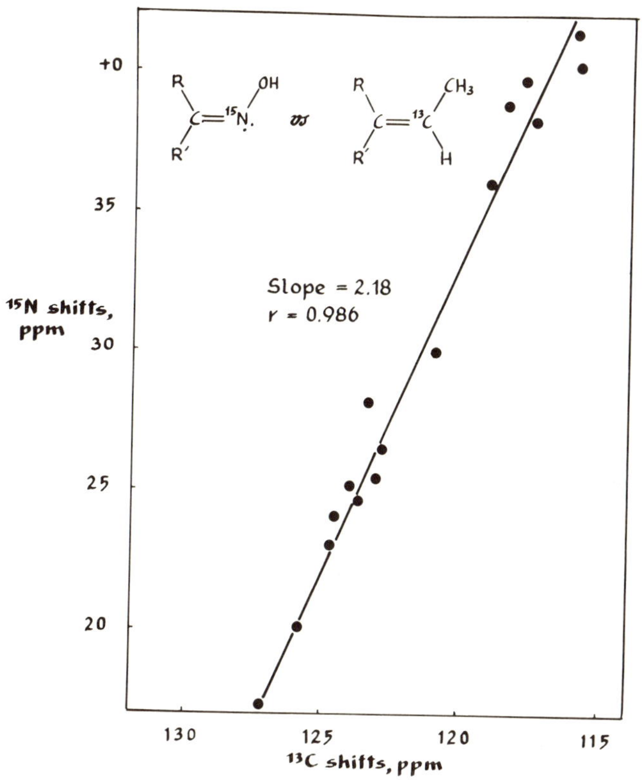

FIG. 19. CORRELATION OF THE ^{15}N SHIFTS of azanols (oximes) with the ^{13}C shifts of methylalkenes.[43]

on the carbon of the C=N bond. Another indication that this is the case is provided by the 4-substituted phenylmethylidenecyclohexanamines (structure **30**) and their conjugate acid salts (structure **31**) with trifluoroethanoic acid. The influence of changes in the X group on the ^{15}N shift of **30** is linear

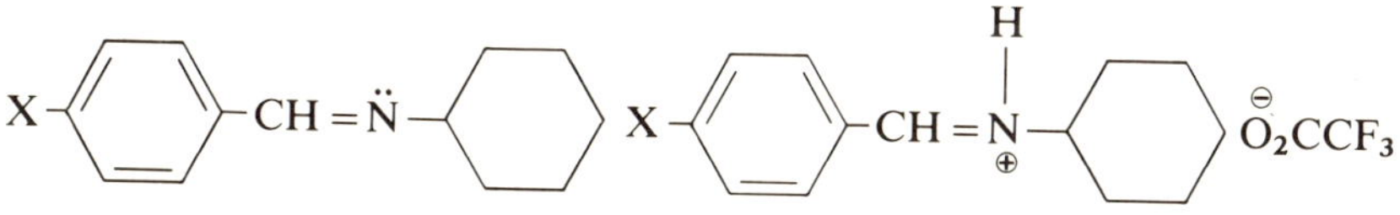

30 **31**

with the Hammett substituent constant of X, which reflects the ability of X to determine the ionization of X-substituted benzenecarboxylic acids (figure 20).[46] When the derivatives of **30** are converted to derivatives of **31**, there is a large upfield shift of about 150 ppm, as expected for quenching of con-

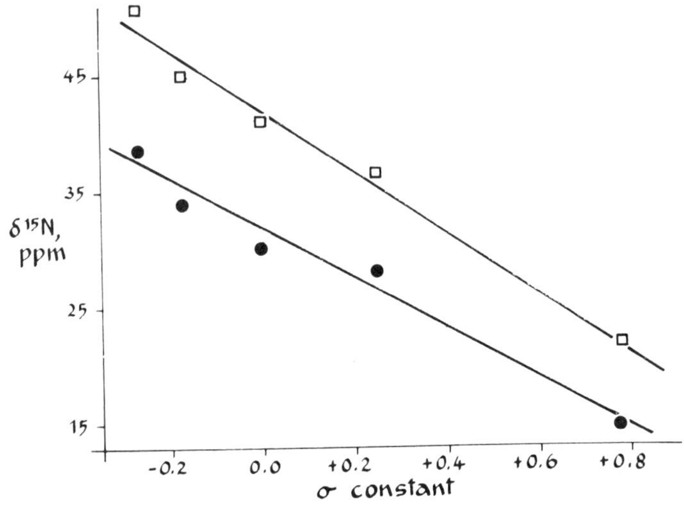

FIG. 20. CORRELATION OF THE ^{15}N SHIFTS of 4-substituted phenylmethylidenecyclohexanamines with Hammett substituent constants (σ). Upper line for methanol solutions, lower line for chloroform solutions.[45]

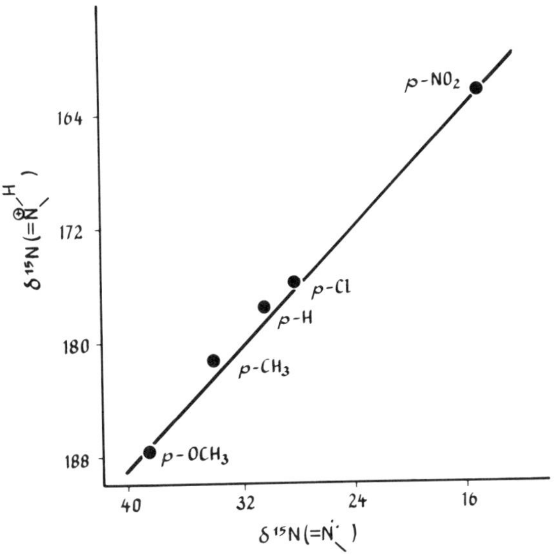

FIG. 21. CORRELATION OF THE ^{15}N SHIFTS of 4-substituted phenylmethylidenecyclohexanamines (structure **30**) and those of their conjugate acids (structure **31**).[46]

tribution of the $n \rightarrow \pi^*$ excited state to the ground state of **30**. The perhaps surprising ancillary result is that the ^{15}N shifts of the derivatives of **30** correlate with those of **31** with a slope of 0.96 and $r = 0.994$ (figure 21). Thus, the effects of the X substituents on the shifts of **30** are virtually the same as for **31**.[47]

Many further applications of ^{15}N NMR to chemistry and biochemistry could be discussed. I will mention here only one possible use of ^{15}N NMR in peptide chemistry, which sheds further light on the importance of the steric effect in influencing ^{15}N chemical shifts. The influence of the saturated R groups of the amino acids alanine, valine, and leucine (parent structure **32**), when combined into di- and tripeptides at the isoelectric point, can be seen from the data given below.[47,48]

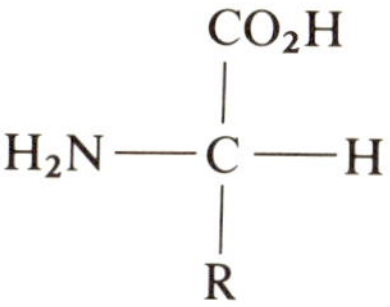

alanine (Ala), R = CH_3
valine (Val), R =
leucine (Leu), R =

32

DIPEPTIDES

$H_3\overset{\oplus}{N}-$		$-\overset{H}{N}-CRHCO_2^{\ominus}$
348.9	Gly - Gly	260.4
335.0	Ala - Gly	260.4
336.8	Leu - Gly	257.1
340.2	Val - Gly	255.8
348.2	Gly - Ala	245.9
335.9	Ala - Ala	246.1
340.6	Val - Ala	242.6
336.7	Leu - Ala	244.0
348.9	Gly - Gly	260.4
348.4	Gly - Ala	245.9
348.6	Gly - Val	252.2
348.9	Gly - Leu	248.1
335.0	Ala - Gly	260.4
335.4	Ala - Ala	246.1
335.1	Ala - Val	252.2
335.1	Ala - Leu	248.7

TRIPEPTIDES

348.9	Gly - Gly - Gly	260.7
	266.7	
347.7	Ala - Gly - Gly	260.7
	266.6	
336.4	Leu - Gly - Gly	260.5
	263.7	
340.1	Val - Gly - Gly	260.5
	262.3	
348.9	Gly - Gly - Ala	246.5
	266.6	
335.4	Ala - Ala - Ala	247.0
	251.8	
349.6	Gly - Ala - Gly	261.7
	252.2	
349.1	Gly - Leu - Gly	259.9
	253.8	

The nitrogen of each amino-acid residue at the amine (left end) terminus of a peptide is seen to have a characteristic shift that is quite independent of the nature of the other amino acids in the chain. For other than *N*-terminal locations in peptide chains, there are three important shift influences. First, the characteristic shift of the nitrogen of the amino acid itself; second, whether it is at the *C*-terminus; and, third, the nature of the amino acid that is bonded to the nitrogen.

It will be seen from the foregoing that ^{15}N shifts (measured in ppm) are very sensitive to structural changes; indeed, roughly twice as sensitive as ^{13}C shifts, and more than an order of magnitude more sensitive than ^{1}H shifts. This makes for the possibility of elucidation of quite subtle intra- and intermolecular interactions by ^{15}N NMR, and the purpose of this paper has been to show something of what we know today of the nature and regularity of these interactions. As we come to understand the scope and physical bases involved, we can expect increasing use of ^{15}N NMR to solve critical chemical problems involving nitrogen compounds.

ACKNOWLEDGMENTS

Supported by the National Science Foundation, and by the Public Health Service, Research Grant No. GM-11072 from the Division of General Medical Sciences.

Contribution No. 5912 from the Gates and Crellin Laboratories of Chemistry, California Institute of Technology, Pasadena, CA 91125.

REFERENCES

1. F. Bloch, W. W. Hansen, and M. Packard, Phys. Rev. **69**, 127 (1946).
2. E. M. Purcell, H. C. Torrey, and R. V. Pound, Phys. Rev. **69**, 37 (1946).
3. See P. C. Lauterbur, Ann. N. Y. Acad. Sci. **70**, 841 (1958).
4. See (a) D. M. Grant and E. G. Paul, J. Am. Chem. Soc. **86**, 2984 (1964); (b) D. K. Dalling and D. M. Grant, ibid. **89**, 6612 (1967).
5. F. J. Weigert, M. Jautelat, and J. D. Roberts, Proc. Natl. Acad. Sci. U.S. **60**, 1152 (1968).
6. See the review by J. D. Roberts, XXIIIrd International Congress of Pure and Applied Chemistry (Butterworths, London, 1971), vol. 7, p. 71.
7. As one indication of the usefulness of the method, the California Institute of Technology has seven NMR spectrometers that are used in one way or the other to take ^{13}C spectra.
8. W. G. Proctor and F. C. Yu, Phys. Rev. **77**, 717 (1950).
9. See M. Witanowski, L. Stefaniak, and H. Januszewski, in M. Witanowski and G. A. Webb, eds., *Nitrogen NMR* (Plenum Press, New York, 1973), chapter 4.
10. J. M. Lehn and J. P. Kintzinger, in M. Witanowski and G. A. Webb, eds., *Nitrogen NMR* (Plenum Press, New York, 1973), chapter 3.

11. J. B. Lambert, G. Binsch, and J. D. Roberts, Proc. Natl. Acad. Sci. U.S. **51**, 735 (1964); G. Binsch, J. B. Lambert, B. W. Roberts, and J. D. Roberts, J. Am. Chem. Soc. **86**, 5564 (1964).
12. Indeed, the first report of a ^{15}N NMR spectrum on enriched ammonium nitrate dates from the time of discovery of the chemical shift. W. G. Proctor and F. C. Yu, Phys. Rev. **77**, 717 (1950).
13. Unpublished work by F. J. Weigert.
14. R. L. Lichter and J. D. Roberts, J. Am. Chem. Soc. **93**, 3200 (1971).
15. D. Gust, R. B. Moon, and J. D. Roberts, Proc. Natl. Acad. Sci. U.S. **72**, 4696 (1975).
16. A. Allerhand, R. F. Childers, and E. Oldfield, J. Magn. Reson. **11**, 272 (1973).
17. R. O. Duthaler, unpublished experiments.
18. J. A. Pople, Proc. R. Soc. London Ser. A, **239**, 550 (1957).
19. G. A. Olah and T. E. Kiovsky, J. Am. Chem. Soc. **90**, 4666 (1968).
20. R. O. Duthaler and J. D. Roberts, J. Am. Chem. Soc. **100**, 3889 (1978).
21. R. O. Duthaler, K. L. Williamson, D. D. Giannini, W. H. Bearden, and J. D. Roberts, J. Am. Chem. Soc. **99**, 8406 (1977).
22. B. F. Spielvogel and J. M. Purser, J. Am. Chem. Soc. **89**, 5294 (1967); J. M. Purser and B. F. Spielvogel, Chem. Commun. **386** (1968); Inorg. Chem. **7**, 2156 (1968).
23. R. L. Lichter and J. D. Roberts, J. Am. Chem. Soc. **94**, 2495 (1972).
24. R. O. Duthaler and J. D. Roberts, J. Am. Chem. Soc. **100**, 3882, 3889 (1978); J. Magn. Reson. **34**, 129 (1979).
25. See ref. 21 for discussion and also one very direct determination of the equilibrium constant for 4, 4-difluoroazacyclohexane, G. Youssif and J. D. Roberts, J. Am. Chem. Soc. **90**, 6428 (1968).
26. R. O. Duthaler and J. D. Roberts, J. Am. Chem. Soc. **100**, 4969 (1978).
27. See H. J. Reich, M. Jautelat, M. T. Messe, F. J. Weigert, and J. D. Roberts, J. Am. Chem. Soc. **91**, 7445 (1969), and references therein to earlier publications by D. M. Grant and co-workers.
28. Unpublished experiments by D. D. Giannini.
29. R. E. Botto and J. D. Roberts, J. Org. Chem. **42**, 2247 (1977).
30. G. R. Sullivan and J. D. Roberts, J. Org. Chem. **42**, 1095 (1977).
31. R. Huisgen, H. Brade, H. Walz, and I. Glogger, Chem. Ber. **90**, 1437 (1957).
32. F. K. Winkler and J. D. Dunitz, J. Mol. Biol. **59**, 169 (1971).
33. K. L. Williamson and J. D. Roberts, J. Am. Chem. Soc. **98**, 5082 (1976).
34. I. Yavari and J. D. Roberts, J. Am. Chem. Soc. **100**, 5217 (1978).
35. I. Yavari and J. D. Roberts, Tetrahedron Lett. #28, 2491 (1978); Biochem. Biophys. Res. Commun. **83**, 635 (1978); Nouveau J. Chimie **3**, 279 (1979).
36. I. Yavari and J. D. Roberts, J. Am. Chem. Soc. **100**, 4662 (1978).
37. H. G. Förster, J. Am. Chem. Soc., in press.
38. R. Hagen, J. P. Warren, D. H. Hunter, and J. D. Roberts, J. Am. Chem. Soc. **95**, 5712 (1973).
39. See H. H. Jaffe and M. Orchin, *Theory and Applications of Ultraviolet Spectroscopy* (New York, John Wiley and Sons, 1962), pp. 186-188 and 364-365.
40. M. Witanowski, J. Am. Chem. Soc. **90**, 5683 (1968), and M. Witanowski, L. Stefaniak, H. Januszewski, and Z. W. Woklowski, Tetrahedron Lett. #20, 1653 (1971).
41. R. L. Lichter and J. D. Roberts, J. Am. Chem. Soc. **93**, 5218 (1971).
42. E. M. Kosower, J. Am. Chem. Soc. **80**, 3253 (1958); E. M. Kosower and G.-S. Wa, ibid. **83**, 3142 (1961).
43. R. E. Botto, P. W. Westerman, and J. D. Roberts, Org. Magn. Reson., **11**, 510 (1978).
44. J. D. Roberts, F. J. Weigert, J. I. Kroschwitz, and H. J. Reich, J. Am. Chem. Soc. **92**, 1338 (1970).

45. P. W. Westerman, R. E. Botto, and J. D. Roberts, J. Org. Chem. **43**, 2590 (1978).
46. R. E. Botto and J. D. Roberts, J. Org. Chem., **44**, 140 (1979).
47. V. Markowski, T. B. Posner, P. Loftus, and J. D. Roberts, Proc. Natl. Acad. Sci. U.S. **74**, 1308 (1977); see also D. Gattegno, G. E. Hawkes, and E. M. Randall, J. Chem. Soc., Perkin Trans. 2, 1527 (1976).
48. T. B. Posner, V. Markowski, P. Loftus, and J. D. Roberts, Chem. Commun. 769 (1975).

FERROMAGNETISM IN TWO DIMENSIONS—SOME MODELS AND APPLICATIONS

by Michael Schick

ABSTRACT

Several physical realizations of two-dimensional models of ferromagnetism are discussed in the context of adsorbed systems. An exact renormalization group treatment of the simplest of these models, the Ising model, is described.

I. INTRODUCTION

I have been interested in phase transitions in lower dimensional systems since studying superconductivity in one dimension as a graduate student of Felix Bloch. This interest has developed at the University of Washington, where I am fortunate to have colleagues expert in both theoretical and experimental Surface Physics. It is in this connection that I have encountered models of ferromagnetism in two dimensions. These models are studied for several reasons. In some cases, the models are intrinsically difficult, such as those that possess a continuous symmetry, and their solution provides insights into other areas encountering similar difficulties. Furthermore, there are interesting experimental systems that can be related in various ways to ferromagnetic models. These systems often provide experimental tests of our understanding of phase transitions that can be obtained nowhere else. Last, there is at least one model that, in addition to being intrinsically difficult and descriptive of real systems, can be solved exactly and thereby serves as a standard against which new theories and methods in Statistical Mechanics can be compared. This is, of course, the two-dimensional Ising model. In what follows, I shall try to illustrate these reasons with a brief discussion of several areas of current activity.

Michael Schick is Professor of Physics at the University of Washington.

II. THE PLANAR ROTATOR OR x-y MODEL

In his famous paper of 1930, "Zur Theorie des Ferromagnetismus,"[1] Bloch showed the existence of spin-wave excitations in an isotropic Heisenberg ferromagnet. The excitation of these waves with temperature leads to a decrease of the magnetization, which, for a three-dimensional simple cubic system at low temperature, can be written

$$M(T,\alpha) \cong N\mu \left[1 - \frac{1}{4\pi^2 K^{3/2}} \int_0^\infty \frac{u^{1/2}du}{e^{u+\alpha} - 1}\right] ,$$

where K is the exchange interaction in units of kT and α is the magnetic energy also in units of kT. In the limit of zero applied field, α vanishes and the integral is simply a constant. The result is the Bloch $T^{3/2}$ law for the decrease of the spontaneous magnetization with temperature. In the two-dimensional Heisenberg model on a quadratic lattice, Bloch noted that the analogous expression is

$$M(T,\alpha) \cong N\mu \left[1 - \frac{1}{2\pi K} \int_0^\infty \frac{du}{e^{u+\alpha} - 1}\right] .$$

In the limit of zero applied field the integral diverges so that "no ferromagnetism exists." Similar arguments were thereafter employed by Peierls[2] to show that there could be no two-dimensional crystalline order at any finite temperature. These results have been strengthened[3] and generalized to other systems. In particular one can show that there exists no Bose-Einstein condensation in two dimensions at any finite temperature.[4] All of these results rest on the existence of a continuous symmetry. The planar model, whose spins have only two components as contrasted to the three components of the Heisenberg spins, is invariant under the continuous symmetry of the two-dimensional rotation group. Thus it has no spontaneous magnetization either.

For years it was inferred from these results that there were no phase transitions in such systems. But in 1966 Stanley and Kaplan[5] found that high temperature series expansions for the two-dimensional planar model indicated a phase transition at which the magnetic susceptibility would diverge at a finite temperature. The low temperature phase of this system was investigated[6] using the kind of spin wave analysis Bloch had introduced in 1930. It was found that at low temperatures the magnetization varied as a temperature-dependent power of the applied field. Because at high temperatures the magnetization and field are linearly related, a phase transition was again indicated. In 1973, Kosterlitz and Thouless,[7] in a remarkable paper, elucidated the nature of this unusual transition. They showed that whereas one kind of excitation, the spin wave, played the role of destroying any spontaneous magnetization, another kind of excitation was responsible for the transition. This excitation is the vortex, a circular pattern of the planar magnetization completely analogous to the vortex in a fluid. At high temperatures such vortices can be created individually, whereas at low temperatures they appear bound to another vortex of opposite circulation. The presence or absence of free vortices substantially alters the properties of the phase. For example, in a two-dimensional fluid, free vortices will cause a flow to dissipate, whereas bound vortex pairs will not. Thus, the system can exhibit superfluidity at low temperatures although there is no Bose condensation. In the two-dimensional solid the analogues of the vortices are dislocations. Free dislocations move under the application of a shear stress and the system behaves like a liquid. At low temperatures the dislocations are bound and the system possesses a finite shear modulus, although there is no long range crystalline order.

The solution of the problem of interacting vortices was carried out by Kosterlitz and Thouless in an approximate way in 1973. The following year Kosterlitz[8] applied Renormalization Group methods.

It is quite in keeping with the spirit of this Festschrift to note that a completely different path to the solution of the planar model taken by Luther and Scalapino[9] employs a method that goes back to another paper by Felix Bloch.

Luther and Scalapino noted that the interacting vortex problem is equivalent to the problem of a one-dimensional Fermi gas with interactions such that electrons can scatter from one side of the Fermi sea to the other. To solve this problem, they employed a method originated by Bloch in 1934 and later developed by Tomonaga.[10] In the paper "Inkohärente Röntgenstreuung in entarteten Fermigas,"[11] Bloch argued that as the incoherent scattering of x rays from a Fermi gas was due to its excitations, which were simply density fluctuations, one should be able to treat this problem in a manner analogous to Rayleigh scattering of light. He therefore replaced the Hamiltonian of the Fermi gas by one that describes its excitations, "soundwaves," as bosons. Proceeding with this hydrodynamic Hamiltonian, Bloch was able to obtain the correct result for the incoherent scattering of radiation from a one-dimensional electron gas. In the three-dimensional case, however, the result was in error by a factor of $\sqrt{3}/2$. This method of converting a one-dimensional Fermi gas into a boson problem was the basis for my thesis, which investigated the possibility of flux quantization in a one-dimensional superconductor. The whole subject of superconductivity in one dimension, raised originally by Little,[12] became very active again a few years ago. Luther and Peschel[13] were able to examine several aspects of this subject by once more employing these methods. Perhaps the greatest utility of the method, however, arises from the fact that many interesting problems, like the planar model, can be mapped onto the one-dimensional electron gas.[14] Recently Luther[15] has shown how to express the three-dimensional Fermi gas in terms of bosons. Presumably he should be able to calculate the incoherent scattering of x rays and emend the errant factor of $\sqrt{3}/2$.

III. MODELS OF TRANSITIONS ON SURFACES

The models discussed above are all characterized by a continuous symmetry, which is related to their lack of spontaneous magnetization. In this section I wish to describe briefly the application of these same models in the presence of various anisotropies that break the continuous symmetry and permit the appearance of a spontaneous magnetization. Consider first the planar model in the presence of an anisotropy that confines the spins to three directions given by the angles 0, $2\pi/3$ and $4\pi/3$. The resulting Hamiltonian is that of the three-state Potts model. (A q-state Potts model describes objects that can exist in q states. If nearest-neighbor objects are in the same state, the energy of interaction is E_1 and is E_2 otherwise. The Ising model is a two-state Potts model.) Clearly there are three degenerate ground states in which the spins all point in the same direction. My interest in this problem is through order-disorder transitions on surfaces that, on first sight, appear to have no relation to ferromagnetism in the presence of anisotropies. But, "Lo and Beware"—certainly my favorite phrase of Felix's—there is such a relation. Consider a substrate that presents a triangular array of adsorption sites, graphite for example. Now adsorb upon it an atom that is somewhat larger than one site, so that it costs energy to put two atoms on adjacent sites. An ordered ground state of such a system is shown in figure 1. A glance at this figure reveals that there are in fact three degenerate ground states, just as in the three-state Potts model. It is argued[16] on the basis of symmetry that the critical properties of this order-disorder transition will be identical to those of the ferromagnetic three-state Potts model. The particular system of helium adsorbed on graphite has been well studied, particularly by my colleagues at the University of Washington, Greg Dash and Oscar Vilches, and their students.[17] Recent measurements by Bretz[18] appear to substantiate the identification above.

Other kinds of order-disorder transitions are related to other models of anisotropic ferromagnets by the same arguments of symmetry. The examples that follow are drawn from a complete list I prepared in collaboration with Eytan Domany, James Walker, and Robert Griffiths.[19]

Consider the planar model in a four-fold anisotropy; i.e., angles 0, $\pi/2$, π, $3\pi/2$. Clearly there are four degenerate ground states. Now consider

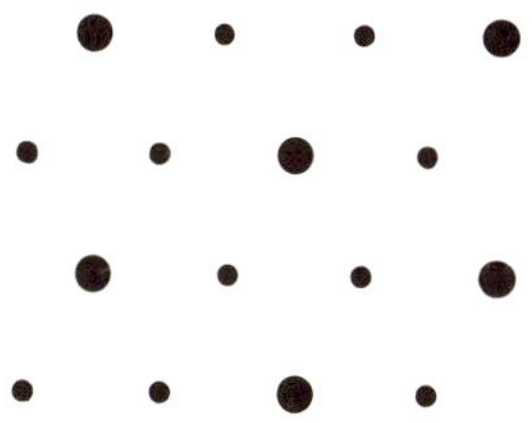

FIG. 1. ONE OF THREE DEGENERATE GROUND STATES of a system of atoms adsorbed on a substrate that presents a triangular array of adsorption sites, and in which the atoms interact via a nearest-neighbor repulsion. Large dots represent occupied sites, small dots unoccupied ones.

FIG. 2. ONE OF FOUR DEGENERATE GROUND STATES that can be realized by a system adsorbed on a substrate that presents a square array of adsorption sites.

the ground state of an adsorbed system shown in figure 2. In such a system there are also four degenerate ground states; two in which columns are occupied and two in which rows are occupied. The symmetries of these two systems are the same and thus their critical properties are predicted to be the same.[20,21]

Other transitions of interest correspond to a two-dimensional Heisenberg model in the presence of anisotropy. Consider first one that causes the spins to align in one of four directions that point to the vertices of a tetrahedron. It is easily seen that the Hamiltonian of this system is just that of the four-state Potts model, because the interaction energy of two adjacent spins $\mathbf{S}_i \cdot \mathbf{S}_j$ is either S^2, if the spins are in the same state, and $-S^2/2$ otherwise. There are four degenerate ground states, as in the planar model with a four-fold anisotropy, but the symmetry relations among these states are clearly quite different. This new symmetry is, however, identical to that of the four degenerate ground states that can be constructed from figure 3a. Thus adsorbed systems that undergo a transition into such an ordered array are expected to display the critical properties of the four-state Potts model. A second anisotropy of interest is the cubic anisotropy, which causes the Heisenberg spins to align along three mutually perpendicular axes. There are six degenerate ground states and the symmetry relations between them are the same as among the six states that can be obtained from figure 3b. The critical properties of these two-dimensional models, the four-state Potts model and the Heisenberg model with cubic anisotropy, are believed to be different from one another and from the models mentioned previously. This belief can be tested by experiments on physical realizations of these models provided by adsorbed systems.

It should not be thought that the transition to just any ordered structure yields different critical behavior. To the contrary, it is predicted

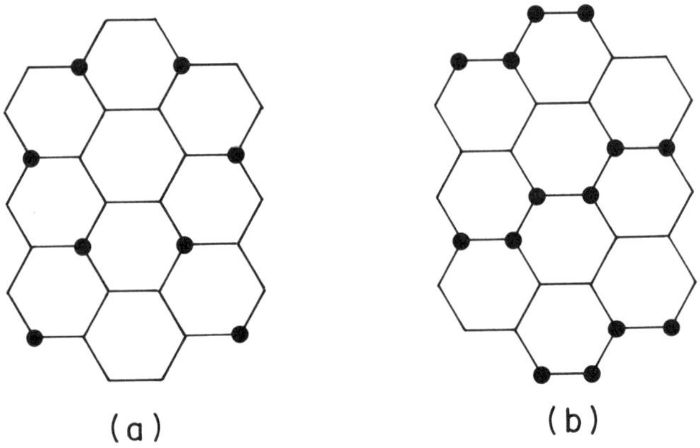

FIG. 3. TWO DIFFERENT KINDS OF GROUND STATES that can be realized by a system adsorbed on a substrate that presents a hexagonal array of sites. Which ground states are realized depends upon the interactions. The degeneracies of the ground states are (a) four, (b) six.

that all critical behavior of such transitions will either be discontinuous or fall into one of a small number of classes, those of the four models above and a few others. One of these is the model of ferromagnetism in the limit of extreme axial anisotropy, the Ising model. The simple symmetry of the two ground states of the ferromagnetic Ising model is shared by several order-disorder transitions. The ground states of two of them are shown in figure 4. These two cases illustrate the idea that different transitions can manifest identical critical properties.

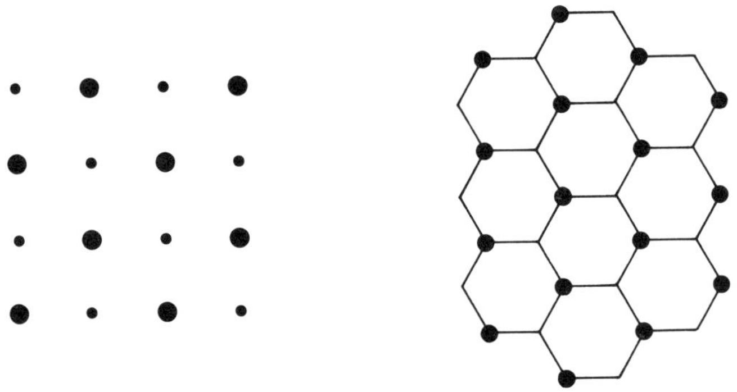

FIG. 4. POSSIBLE GROUND STATES on square or hexagonal arrays of adsorption sites. Each ground state is doubly degenerate.

IV. THE ISING MODEL AND ITS INFINITESIMAL RENORMALIZATION GROUP TRANSFORMATION

My interest in the Ising Model stems from the desire to describe the ordering transition shown by helium adsorbed on graphite, which was discussed above. An obvious Hamiltonian to describe this system is that of a lattice gas in which an atom can exist only on adsorption sites and nowhere else. As a site is either occupied or it is not, there is clearly a simple mapping from this Hamiltonian to that of an Ising model in which a spin at each site is either up or down. The repulsive interaction between atoms translates into an antiferromagnetic interaction between spins. This antiferromagnetic Ising model is, like the ferromagnetic one, exactly soluble only in zero magnetic field, which corresponds to zero magnetization. In the lattice gas analogy this corresponds to the experimentally inaccessible region of a gas density of one atom for every two adsorption sites. At the time at which the experiments on helium were performed that measured the variation of the transition temperature with coverage, there were no approximate solutions of the model described above that predicted this quantity. The few mean-field calculations that had been performed gave incorrect results at the coverage of one-half where the exact result was known. I decided to learn something about the Renormalization Group ideas of Ken Wilson[22] by applying them to this problem.[23] Let me recall a few of the relevant ideas here.

In principle, the aim of a calculation in Statistical Mechanics is to obtain the partition function

$$Z = \sum_{S} e^{-H(\{K\},S)} \quad , \tag{1}$$

where for definiteness I consider the Hamiltonian of a set of spins S with interactions $\{K\}$. The factor of $1/kT$ is absorbed into the Hamiltonian. On calculating the partition function, one knows the free energy and all its singularities. Unfortunately, the trace in Eq. (1) can rarely be performed. Traditional mean-field methods for approximating the sum yield singularities in the approximate free energy that differ significantly from the true singularities. The idea of the Renormalization Group (RG) is to carry out the infinite reduction of the degrees of freedom in Eq. (1) in an infinite number of identical steps. Thus in each step the system of spins S is mapped onto another system of fewer spins S' by means of a projection operator $P(S',S)$; i.e., Eq. (1) is rewritten

$$Z = \sum_{S'} \sum_{S} P(S',S) e^{-H(\{K\},S)} \tag{2}$$

The partition function is unchanged provided that the projection operator satisfies

$$\sum_{S'} P(S',S) = 1 \quad . \tag{3}$$

The sum over spin S configurations in Eq. (2) produces an operator that depends only on the spins S'. A renormalized Hamiltonian $H'(\{K'\},S')$ that has the same functional form as $H(\{K\},S)$, but in which the interaction strengths are renormalized, is defined by

$$e^{-H'(\{K'\},S')} = \sum_{S} P(S',S)\, e^{-H(\{K\},S)} \quad , \tag{4}$$

so that Eq. (2) takes the form

$$Z = \sum_{S'} e^{-H'(\{K'\},S')} \quad , \tag{5}$$

analogous to Eq. (1). Thus, starting with H, we obtain a renormalized Hamiltonian H', which has fewer degrees of freedom. We can apply the projection operator again and obtain the entire trace by iteration. The transformation on the Hamiltonians given by Eq. (4) can also be viewed as a transformation on the set of interactions $\{K\}$ and be written symbolically as

$$\{K'\} = R\{K\} \tag{6}$$

where R is the RG (renormalization group) operator. As the entire trace emerges from repeated application of Eq. (6), these recursion relations contain all the information about the singularities of the partition function. The methods for extracting this information are well known.[22] In particular, phase transitions are associated with fixed point solutions $\{K^*\}$ of the nonlinear recursion relations defined by $\{K^*\} = R\{K^*\}$. The critical exponents characterizing the singularities at the transition are obtained from the eigenvalues of the operator R linearized about the fixed point. It is characteristic of RG methods that by focusing on the recursion relations, Eq. (6), they produce information about critical properties first. Other information, such as the behavior of the thermodynamic functions away from the critical region, emerges later.

The great advantage of the RG approach is that the non-analyticities in the free energy emerge from the infinite number of applications of the RG transformation. The recursion relations themselves, however, are in general analytic. As such they can, if necessary, be more safely approximated without doing great violence to the non-analyticities in the partition function.

A particular RG transformation is determined by specifying the projection operator $P(S', S)$. As an example of a projection operator that maps one spin system in position space onto another, I shall employ one drawn from the original work of 'Dorus Niemeyer and Hans van Leeuwen, who introduced these real space methods.[24]

Consider an Ising model on a triangular lattice and group all the spins into cells as shown in figure 5. Each spin is assigned to only one cell. The projection operator $P(S', S)$ is written as a product over all cells

$$P(S', S) = \prod_i p\,(S_1^i, S_2^i, S_3^i; S_i') \tag{7}$$

where the cell projection operator is given by

$$p(S_1, S_2, S_3; S') = \frac{1}{2}[1 + \frac{1}{2}(S_1 + S_2 + S_3 - S_1S_2S_3)S'] \; . \tag{8}$$

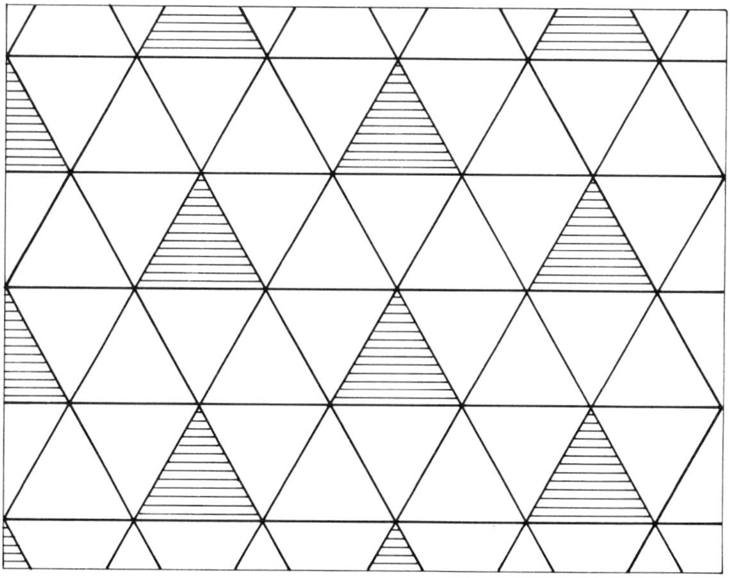

FIG. 5. AN ASSIGNMENT OF SPINS TO CELLS. The cells are shown shaded.

It is easily seen that p maps any configuration of a cell in which a majority of spins are up to the state in which S' is up. As there is one cell spin S' for each three site spins S, the Hamiltonian H' will contain one-third as many degrees of freedom as H. Therefore, the recursion relations Eq. (6) will be finite difference equations. Unfortunately, when Eqs. (7) and (8) are substituted into Eq. (4), the sum can not be carried out, and approximations are necessary.[25] As noted earlier, however, the entire procedure ensures that the results of these approximations will, in general, be superior to mean-field methods.

In 1977–78 I spent six months visiting the Technische Hogeschool in Delft, Holland, in order to improve my understanding of the RG via discussions with Hans van Leeuwen and Henk Hilhorst. For all of the considerable success of the position space RG methods, Hans was dissatisfied that, by their very nature, the cell-spin approach decreased the degrees of freedom by a finite amount and led to difference equations. Ideally, one would like to perform an infinitesimal transformation. In such a case the content of the recursion relations would be expressed in the canonical language of theoretical physics, partial differential equations. Such infinitesimal transformations can be carried out in principle by working in momentum space, i.e., with a Hamiltonian expressed in terms of spin Fourier components. Traces are then integrations in momentum space, and, by integrating over an infinitesimal shell of momenta, one can reduce the degrees of freedom infinitesimally. In practice one can not carry out the integration explicitly and approximation methods must be employed.

Henk and Hans and I were able to formulate an infinitesimal RG transformation in real space by proceeding as follows: We considered a two-dimensional triangular lattice of Ising spins, which has a lattice constant a and the shape of an equilateral triangle with side of length L (the unprimed system). This system is mapped onto a similar triangular lattice also of length L but with one fewer lattice site along each edge (the primed system). The lattice constant of the primed system is therefore larger than that of the unprimed one by the fractional amount $\delta a/a = a/L - a$. The number of spins in the two systems differs by a fraction of order a/L, which is infinitesimal in the thermodynamic limit. Figure 6 shows the primed and unprimed system denoted by crosses and circles respectively. We found it convenient to rewrite the basic RG equations, (3) and (4), by introducing the combined Hamiltonian of the S and S' system defined by

$$e^{-H_c(S',S)} \equiv P(S',S)e^{-H(S)}$$

Then Eqs. (3) and (4) can be rewritten in the more symmetric form

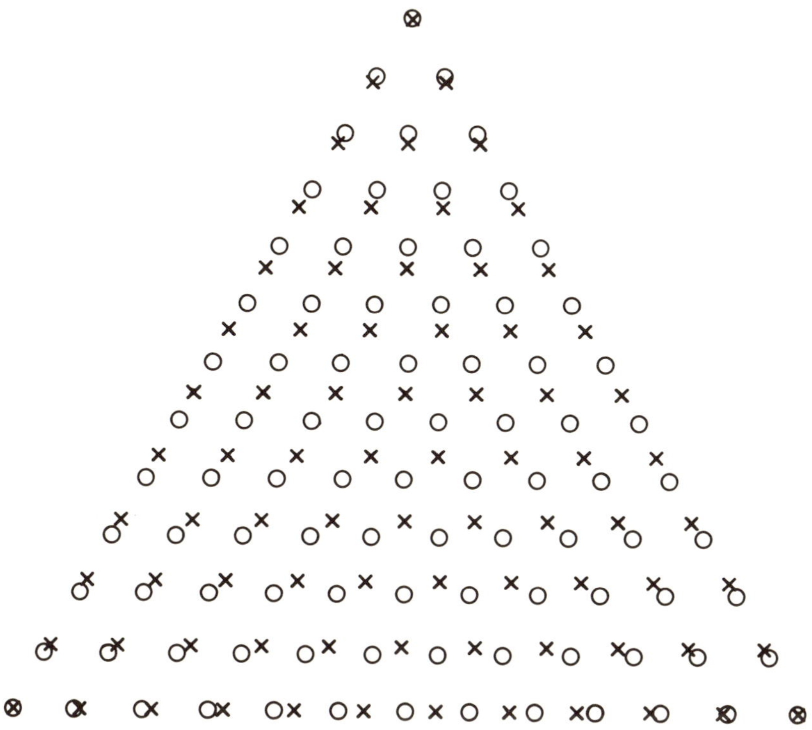

FIG. 6. THE TRIANGULAR SYSTEM OF OPEN CIRCLES is mapped to the triangular system of crosses by the Renormalization Group transformation.

$$e^{-H(S)} = \sum_{S'} e^{-H_c(S',S)} , \tag{9}$$

$$e^{-H'(S')} = \sum_{S} e^{-H_c(S',S)} . \tag{10}$$

The form of $H_c(S',S)$ was at our disposal and we chose a local Hamiltonian that couples nearby spins S and S'. Intuitively we expected that pairs of primed and unprimed spins that are quite close to one another, such as occur near the vertices in figure 6, should be coupled more strongly by H_c than pairs that are not so close as in the center. Thus we expected the coupling constants in H_c to be spatially dependent.

From Eqs. (9) and (10) it is clear that if $H_c(S',S)$ were symmetric in S and S' then H' would equal H and the RG transformation would reduce to the identity. However, if H_c fails infinitesimally from being symmetric, it will generate an infinitesimal difference between H and H' as desired. It turns out that the infinitesimal deviation from symmetry that is needed is provided precisely by the spatial gradients of the coupling constants. When the couplings vary appreciably only over distances of the order L, then $H' - H$ is of the order a/L and therefore infinitesimal in the limit of infinite L. The spatial dependence of H_c led us to consider spatially dependent Hamiltonians H and H'. We all vigorously resisted being led down such a path. It is one thing to consider one-particle interactions that vary with position, as in the case of an inhomogeneous applied magnetic field. It is quite another to consider two-particle interactions that so vary! When we finally yielded, however, we found to our delight that we were able to carry out all operations explicitly and thereby obtain exact RG equations for the Ising model with nearest-neighbor interactions. As we had desired, our recursion relations were non-linear first-order partial differential equations. They appear to be too complicated to solve in general, but we were able to locate the fixed point governing the phase transition. We then linearized the recursion relations about this fixed point and obtained the eigenvalue that demonstrates the logarithmically singular specific heat.[26]

Henk, Hans, and I spent a morning in Leiden talking with mathematicians about our non-linear partial differential equation. They informed us that little was known about this kind of equation. One thing could be said, however. The solutions of such equations often tended to develop very abrupt changes over very short distances. I smiled as I mused to myself that such solutions of non-linear equations are very "hot" currently. They go by the name of "solitons." The simplest example of a soliton that I know of occurs in ferromagnetic systems. It is called a Bloch wall.[27]

ACKNOWLEDGMENT

This work was supported by National Science Foundation Grant DMR 7920785.

REFERENCES CITED

1. F. Bloch, Z. Phys. **61**, 206 (1930).
2. R. E. Peierls, Ann. inst. Henri Poincaré **5**, 1771 (1935).
3. N. D. Mermin and H. Wagner, Phys. Rev. Lett. **17**, 1136 (1966).
4. P. C. Hohenberg, Phys. Rev. **158**, 383 (1967).
5. H. E. Stanley and T. A. Kaplan, Phys. Rev. Lett. **17**, 913 (1966).

6. F. J. Wegner, Z. Phys. **206**, 465 (1967); V. L. Berezinskii, Sov. Phys.—JETP **32**, 493 (1971).
7. J. M. Kosterlitz and D. J. Thouless, J. Phys. **C6**, 1181 (1973).
8. J. M. Kosterlitz, J. Phys. **C7**, 1046 (1974).
9. A. Luther and D. J. Scalapino, Phys. Rev. **B16**, 1153 (1977).
10. S. Tomonaga, Progr. Theor. Phys. **5**, 544 (1950).
11. F. Bloch, Helv. Phys. Acta **7**, 385 (1934).
12. W. A. Little, Phys. Rev. **134** A, 1416 (1964).
13. A. Luther and I. Peschel, Phys. Rev. **B9**, 2911 (1974).
14. Other examples can be found in A. Luther and I. Peschel, Phys. Rev. **B12**, 3908 (1975).
15. A. Luther, Phys. Rev. **B19**, 320 (1979).
16. S. Alexander, Phys. Lett. **A54**, 353 (1975).
17. M. Bretz, J. G. Dash, D. C. Hinkernell, E. O. McLean, and O. E. Vilches, Phys. Rev. **A8**, 1589 (1973).
18. M. Bretz, Phys. Rev. Lett. **38**, 501 (1977).
19. E. Domany, M. Schick, and J. S. Walker, Phys. Rev. Lett. **38**, 1148 (1977); E. Domany, M. Schick, J. S. Walker, and R. B. Griffiths, Phys. Rev. **B18**, 2209 (1978); E. Domany and M. Schick, Phys. Rev. **B20**, 3828 (1979).
20. S. Krinsky and D. Mukamel, Phys. Rev. **B16**, 2313 (1977); E. Domany and E. K. Riedel, Phys. Rev. Lett. **40**, 561 (1978).
21. J. V. Jose, L. P. Kadanoff, S. Kirkpatrick, and R. D. Nelson, Phys. Rev. **B16**, 1217 (1977).
22. I think these ideas are most clearly expressed in K. G. Wilson, Phys. Rev. **B4**, 3174 (1971). There exist numerous excellent reviews; vol. 6 of *Phase Transitions and Critical Phenomena*, edited by C. Domb and M. S. Green (Academic, London, 1976), is also an excellent source.
23. The result of this study is M. Schick, J. S. Walker, and M. Wortis, Phys. Rev. **B16**, 2205 (1977).
24. Th. Niemeyer and J. M. J. van Leeuwen, Physica **71**, 17 (1974).
25. There are some simple one-dimensional cases for which the trace can be carried out explicitly; these are particularly instructive. See, for example, D. R. Nelson and M. E. Fisher, Ann. Phys. (N.Y.) **91**, 226 (1975). Previous to the calculation reported in the text, only simple systems such as these had been solved exactly by RG methods.
26. Our results and the details of the calculations are found in H. J. Hilhorst, M. Schick, and J. M. J. van Leeuwen, Phys. Rev. Lett. **40**, 1605 (1978), and Phys. Rev. **B19**, 2749 (1979).
27. F. Bloch, Z. Phys. **74**, 295 (1932).

TEN YEARS OF NEUTRON PHYSICS WITH FELIX BLOCH AT STANFORD 1938–1949

by Hans H. Staub

ABSTRACT

Work on the magnetic properties of the neutron occupied Felix Bloch and his colleagues during the pre-World War II years at Stanford. This paper gives a personal view of that work and describes the construction and use of the 27-3/4 inch cyclotron at Stanford by Bloch, Bradbury, Staub, and Stephens.

I met Felix Bloch for the first time when, as an assistant of W. Pauli, he had to correct and grade the weekly homework papers of the physics and mathematics students taking Pauli's lectures on classical theoretical physics. I do not think that Felix remembers my papers. They were certainly not brilliant but fortunately also not so catastrophically bad as to be easily remembered. While he was at Leipzig with Heisenberg and from 1933 on at Stanford, he quite often came to Zürich, and since only about two dozen physicists were at the ETH and the University of Zürich, we often met in colloquia, in seminars, and informally at some of the physics professors' homes. In 1937 I received an international student fellowship and a stipend from the ETH in Zürich to work at C. C. Lauritsen's Kellogg Radiation Laboratory at Cal Tech. On our way to Pasadena, by boat and train of course, my wife and I stayed for a few days in New York, where we did all the usual sight-seeing. One evening, while we were wandering in the enormous crowd on Times Square, my wife suddenly exclaimed, "Wasn't this Felix Bloch who just passed us?" Sure enough, he too had recognized us and turned around. He was on his way from California to Zürich for a visit to his parents and had taken an evening stroll with Melba Phillips. He

Hans Staub is Professor of Physics at the University of Zürich.

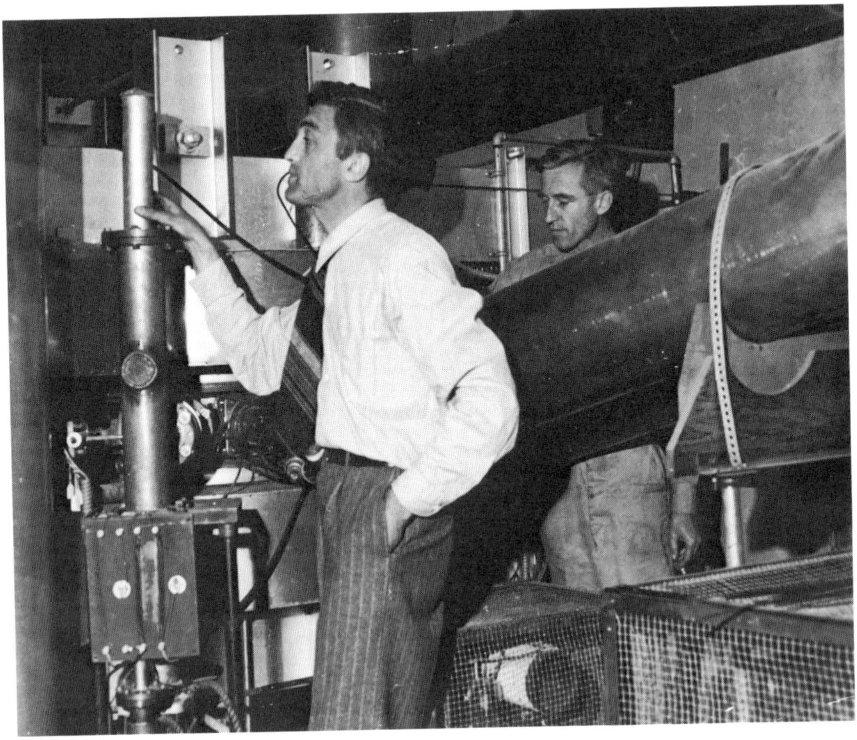

FELIX BLOCH, standing at the dee support end of the Stanford cyclotron in 1946. The author (H. Staub) is in the background.

was greatly surprised that we took the same stroll in the opposite direction. The four of us had a merry evening. Felix took us greenhorns up to the sky-room of the Rockefeller Center. He also kindly took the key of my father's house, which by mistake I had taken along, back to my parents.

More than half a year later, when I started worrying over future employment, either in the USA or in Switzerland, I got a letter from Felix asking whether I would be interested in a position as instructor of physics at Stanford. He wanted very strongly to have the position filled with an experimentalist in nuclear physics, since together with Norris Bradbury and Howard Tatel he had just started the attempt to measure the magnetic moment of the neutron. I certainly was very much interested and with my wife and our first baby we drove up to Stanford, where I had to deliver the usual trial talk in the journal club. Apparently it was successful, and in September 1938 we moved to Stanford, where an unforgettable ten years of collaboration with Felix began.

Our first project was to obtain a sufficiently powerful source of neutrons. It was quite clear that the first pathetic experiments using Ra-Be sources for neutrons could not succeed for intensity reasons, although the temporal constancy of intensity was a great advantage. Keeping in mind that even a few hundred dollars were quite hard to pry loose from the department budget, we decided to build a d-d accelerator to be operated on the 170 KV voltage source that had been in existence at Stanford for some time and that was used for the famous x ray work of Webster, Kirkpatrick, and Hansen. For the construction of the accelerator tube I had quite a bit of experience from Zürich and Pasadena. The machine, practically all home made, was completed in the summer of 1939, but it was never used for work on the magnetic moment of the neutron. While the machine was still under construction, Felix got quite unexpectedly the opportunity to do the first successful magnetic resonance experiment with Luis Alvarez at the Berkeley Radiation Laboratory's 37″ cyclotron. After completion, however, the Stanford machine was a suitable source of energetic neutrons for work on resonance scattering of neutrons on helium, which William Stephens and I had started at Pasadena and which was completed at Stanford together with Howard Tatel.[1] In this work Felix too participated by working out the theory of this particular type of resonance scattering.[2] Thus we were able to compare the measured maximum cross-section with the theory. Our measurements showed some evidence of a splitting of the P 3/2 and P 1/2 levels of a few hundreds KeV, but the confidence level was quite low. Indeed, it turned out later that the splitting was much larger. Our measurement of the cross-section, however, and this is the important point, was on an absolute scale (via comparison with the n-d and n-p cross-sections) and allowed therefore a comparison with Bloch's theoretical results without any open normalizing factor. The magnitude of the cross-section of the levels decreases, of course, rapidly with increasing splitting. Consequently the observed maximum cross-section of 0.7 barns showed that the splitting was at least somewhat larger than the width.

The work on the neutron magnetic moment with Alvarez had convinced Bloch that in order to do a more accurate experiment, it was necessary to study first several related problems, such as the polarization of neutrons by magnetic scattering and also the problem of the direct comparison with the proton magnetic moment. For all this work we simply needed a still more intense slow neutron source. Felix's mind was now dead set on getting a cyclotron for accelerating deuterons to some MeV with currents of some tens of microamps on a beryllium target. Such a device would be equivalent to a source of Ra-Be of some kilograms. I too shared his desire, but I was appalled to think of the financial consequences. We started, together with Norris Bradbury, and later W. E. Stephens, to do some serious engineering considerations with a few firm principles in mind: Let us not go into any fancy development work; from Stanford to Berkeley it is only 30 miles to get competent advice; let us concentrate on a modest machine and avoid unnecessary changes in order to lower the cost so that the project can be realized with a few thousand dollars.

From then on things moved very quickly. Felix proved himself to be not only an outstanding theoretician of magnetism, but also a very able constructor of a large magnet. Moreover, he was a wizard in raising the four thousand dollars needed for the machine. Let me quote a footnote in the first of the many papers that resulted from the work with this machine[3]: "The construction of this 27-3/4″ cyclotron by Bloch, Bradbury, Staub, and Stephens was made possible by grants from the Rockefeller Foundation, the Research Corporation and from various private sources. It was completed in the autumn of 1941, giving under normal operation a beam of about 10 μamp of 2.5 MeV deuterons." We intended to publish a short account on the peculiar features of this machine, but like many other projects, our intention fell victim to heavy war-time activities. It seems therefore fitting to use this occasion for an account.

The magnet was built from stacks of 1-inch mild steel plates, machined only in places where it was necessary for stability and continuity of the flux. For economic reasons we decided to use forced air cooling, although this resulted in a relatively small power capacity and a relatively weak field of about 12 KG. By minimizing the costs of the magnet, we found a curious relation for the cost of the copper and iron. At the prevailing price level it turned out that these two amounts should be about equal. The magnet was energized by a 10 KW AC-DC converter generator, which the University owned as an emergency standby. Felix succeeded in extracting the machine from the office of buildings and grounds, provided it would be available to them at any time. To meet this condition we installed an immense antiquated double pole-double throw switch, but it was never used.

A particularly improbable incident just after completion of the machine remains unforgettable to me. It was on the last day of the spring quarter, the machine had been assembled to the last bolt, and we were ready to start testing and shimming the magnet. We felt that we had done a lot of hard work and deserved a few days of vacation, hiking in the Sierras. We still had some time left that afternoon, and so we picked, more or less at random, two equal circular iron plates from a stack that had been prepared for systematic shimming. To our utter surprise we got a beam of several microamps. Happily we shut down and went on our vacation. Upon returning a few days later we started a systematic shimming procedure. The two haphazardly chosen plates were taken out and replaced pairwise by slightly different ones. But whatever we did, even inserting the two former plates, we could not get even an indication of a beam. After two days of trying, we sat crestfallen around the machine until somebody had the bright idea to have a good look at the two original plates. Sure enough, the two differed in thickness by less than 1/64 of an inch, and because of a slight mechanical asymmetry of the magnet, the thicker one had to be on the bottom. But this was accidentally just the case in the first try, and it was not even the sole incredible accident. It also turned out later that their size was exactly optimal.

The 10 KW, 10 Mc oscillator was of the conventional single tube tuned plate, tuned grid type, the tube being a demountable water cooled triode attached to a small oil diffusion pump. It had been built by Litton for a charity prize. The grid circuit consisted of a 30-cm diameter coaxial line of 10 m length. The low voltage arc ion source, which initially gave a deflected deuteron beam of about 10 μamps, was later slightly modified, thereby increasing the current to 40 μamps.

Concurrently with the construction of the cyclotron, we had to build all the clumsy electronic and neutron counters, which also required much room and power, by ourselves. The idea of having a technician or a machinist working full-time on our experiments never even entered our minds. Felix was a marvelous collaborator. Within a short time he became quite an expert in electronics and enjoyed these simple techniques. When a circuit he had built with his own hands worked just the way it was expected to, he was as pleased as a child with a toy. One incident was quite typical of his way of thinking in simple models and with uncomplicated conclusions. We had a problem concerning the shape of the undershoot from an RC coupled pulse amplifier. I told Felix that whatever magnitude R and C had, the area of undershoot would be the same as that of the upshoot. For an instant he seemed puzzled and asked how one could make such a general statement. When I replied that the output signal could not contain a DC component, he was delighted by the simplicity of this explanation.

It was also typical of Bloch that after the cyclotron was successfully operating, he would not attack the main experiment, the precision determination of the neutron magnetic moment, without thorough preparation. In order to do a thorough experiment, we had to get not only an intense slow neutron beam but even more important, a highly polarized one. Bloch himself had shown long before[4] that by magnetic scattering in ferromagnetic substances a beam of neutrons passing through magnetized iron would be appreciably polarized. But it was not understood why in a ferromagnet the deviation from complete saturation had to be so small. In fact, Bloch and Alvarez in their work had to use rather large electromagnets to magnetize the soft iron polarizer and analyzer plates in order to get a 6% single transmission effect from a 4-cm Swedish iron plate. Just about the time we started in 1941, Halpern and Holstein[5] showed that the high degree of saturation in a polycrystalline medium was necessary to avoid depolarization of the neutrons in the boundary zones of the microcrystals. The first work with our machine by Bloch, Hamermesh, and myself[3] was a thorough study of the so-called single transmission effect, which gave a surprisingly good verification of the Halpern-Holstein theory and reasonable and quite accurate values for the magnetic scattering cross section and the size of the microcrystals. However, the result for the magnetic scattering cross section was found to be considerably larger than the value calculated by Hamermesh.[6] Since the first experimental data were taken with neutrons emerging from a block of paraffin of rather complex form, it was felt that single transmission measurements should be conducted with monochromatic thermal neutrons. For this purpose the cyclotron was pulsed and a time-of-flight arrangement, which shortly afterward came in very handy, was built by E. Fryer.[7] His results, however, were still higher than the theoretical value by about the same factor as before.

Any further work on the magnetic properties of the neutron came to an abrupt end in midsummer of 1942, when Felix and I, with the collaboration of M. Hamermesh and D. B. Nicodemus and others, took over a contract with the Manhattan District project prior to the formation of the Los Alamos Laboratory. Our task was the determination of the spectral distribution of the neutrons from fission induced by thermal neutrons in U 235. At that time this was quite a formidable experiment, since no enriched uranium was available. A direct spectral distribution measurement was out of the question, since even the most sophisticated moderator arrangement would still bring a large number of fast neutrons from the source into the detector and drown the few fission neutrons of about the same energy from a fission source of a few kg of natural uranium metal. The difficulties were overcome by two important experimental novelties. The first gave a purely thermal atmosphere of fission-inducing neutrons by capturing a burst of fast neutrons from the cyclotron in a cubic cavity of about 1 m^3 whose walls consisted of highly purified reactor graphite, which had just become available. The thermal neutrons had a lifetime of about 2 millisec., so that after about one millisec. none of the fast neutrons but practically all of the slow ones were present. The second trick concerned the recording sensitivity of the hydrogen recoil pulse ionization chamber. The primary fast neutron pulse put an enormous charge on the collecting electrode and the grid of the first tube connected to it, and would thereby paralyze the counter arrangement with its relatively large time constant for a considerable time. The difficulty was overcome by compensating this charge through a time modulated negative feedback arrangement. The result we obtained was quite good.[8] It differed from later precise measurements mainly by a 30% lower intensity of fast neutrons around 2 to 3 MeV. As we had suspected, this difference was due to an appreciable inelastic scattering of the fission neutrons in the large mass of uranium and mainly in the very thick walls of the ionization chamber container, whose dimensions had to conform with the California State safety regulations.

By the end of June 1943 we had completed this work and moved with some of the equipment to Los Alamos. A few months later Felix joined the radiation laboratory at Cambridge, Massachusetts, and our collaboration was interrupted for more than two years.

In September 1945, I returned for a few days' visit to Stanford. Felix had already been back for some time and had started peaceful work with great enthusiasm. I remember very well how, on a sunny afternoon in his garden, he told me that he believed he had found the really decisive method for comparing the magnetic moments of the neutron and proton. In his familiar and simple way, using purely classical ideas and models, he explained his experiment, and I heard for the first time about nuclear induction. He urged me to return immediately to Stanford to participate in this work, but much as I regretted it (and I still do), I had to return to Los Alamos until February 1946 for a lot of clean-up work. When I finally returned, Felix, Bill Hansen, and Martin Packard had just completed successfully the first nuclear induction experiments, using the same old 3-inch lecture demonstration magnet we had used for the neutron polarization experiments, whenever it was not needed for lecture demonstrations.

Fortunately the times of extreme poverty in pure physics were over by then, and although Felix and I still practiced quite a bit of sporty austerity, ONR and AEC grants gave one the comfortable feeling that the procurement of a modest magnet to provide a 10 KG field over some thousand c.c.'s no longer meant an uncertain and dangerous financial venture.

Thus, in a carefree atmosphere, our reassembled group of Bloch, Nicodemus, and myself started the precision measurement of the neutron magnetic moment, carefully preparing and checking each step without hurry. First of all, we repeated the pre-war experiments on neutron polarization in magnetized iron, extending the measurements to magnetizing fields of 10,000 Oersted. We were rather pleased to find that the saturation polarization was considerably larger than the previous value obtained from a large extrapolation. For a 3/4-inch thick iron plate the saturation value of the single transmission effect was almost 7%, corresponding to a polarization of the neutrons of 35%. This increase meant that the observable resonance would give a comfortable transmission effect with a maximum of about 15%.

Proton nuclear induction was used not only for the actual measurement of the magnetic moment ratio, but for the first time also for stabilization of the magnetic field.

In a little less than two years, between 1946 and 1948, the work was completed in all details.[9] The accuracy of the result was not improved by an order of magnitude until more than 25 years later. During the last two years of my activity at Stanford, Felix and I each directed separate projects in the field of nuclear magnetic moments, but we still cooperated a great deal, discussed our problems, and helped each other. With J. Fleeman and D. B.

Nicodemus, I did a final series of thorough experiments on neutron polarization, which cleared away the few remaining problems on polarization by magnetic scattering.[10]

With E. Rogers,[11] we did what we called an amusing but perhaps unnecessary experiment. We determined the signs of proton and neutron magnetic moments directly by using a true rotating field instead of an oscillating one. I remember vividly how pleased Felix and also I. I. Rabi were, that the transition experiment was done for once with a true rotating field. At that time, Bloch had just started with C. Jeffries the measurement of the value of the proton magnetic moment in terms of the nuclear magneton.[12] They used the ingenious method of the anticyclotron resonance, and I was happy that particularly during Felix's absence in Europe, I could give a hand in the instrumentation of that beautiful experiment.

Then, in September 1949, our collaboration of ten years came to an end, when I accepted the position of the chair of experimental physics at the University of Zürich.

I have often been asked why I felt that doing physics with Felix led to such a particularly agreeable and fruitful partnership, since after all, theoreticians and experimentalists also collaborate successfully in many other places. But I am convinced that the reason for this particularly productive collaboration lies in the fact that Felix's role was much more than that of a pure theoretician, who advises the experimenters, tells them what to do but not how, and possibly participates in the evaluation of data. Felix is simply just a physicist, recognizing the importance of both theory and experiment and acting correspondingly by participating in the combined activity. But likewise—and this again is a very positive asset—he expects justly that his collaborators understand the theory pertaining to their experiments quite thoroughly. This gift of looking at physics as a unity has unfortunately nowadays become rather rare.

REFERENCES

1. H. Staub and H. Tatel, Phys. Rev. **58**, 820 (1940).
2. F. Bloch, Phys. Rev. **58**, 829 (1940).
3. F. Bloch, M. Hamermesh, and H. Staub, Phys. Rev. **64**, 47 (1943).
4. F. Bloch, Phys. Rev. **51**, 994 (1937).
5. O. Halpern and T. Holstein, Phys. Rev. **59**, 960 (1941).
6. M. Hamermesh, Phys. Rev. **61**, 17 (1942).
7. E. M. Fryer, Phys. Rev. **70**, 235 (1946).
8. D. B. Nicodemus and H. H. Staub, Phys. Rev. **89**, 1288 (1953).
9. F. Bloch, D. Nicodemus, and H. H. Staub, Phys. Rev. **74**, 1025 (1948).
10. J. Fleeman, D. B. Nicodemus, and H. H. Staub, Phys. Rev. **76**, 1774 (1949).
11. E. H. Rogers and H. H. Staub, Phys. Rev. **76**, 980 (1949).
12. C. D. Jeffries, Phys. Rev. **81**, 1040 (1951).

"DOMAIN WALLS" IN MAGNETIC SUPERCONDUCTORS

by H. Suhl

ABSTRACT

A Landau Ginsburg theory is presented for the apparent first order transitions in the "magnetic superconductors" of the type $ErRh_4B_4$. The magnetic and superconducting order parameters are in competition because of both paramagnetic and diamagnetic coupling. I discuss the extreme cases in which one or the other coupling dominates.

I. INTRODUCTION

One of the many problems in physics clarified by Felix Bloch is that of the spatial variation of magnetization in ferromagnets. In the materials available at the time, he advanced the notion of what is now known as the Bloch wall, in which the only spatial changes that could be expected were changes in direction, not in magnitude, of the magnetization. A relatively slow turning of the magnetization, extending over a hunded angstroms or so, does not require much energy, about as much as the electromagnetic energy residing in fringing fields. Thus the introduction of these walls will sometimes lower the total energy because, suitably disposed, they inexpensively lower the magnetic energy stored in the fringing field. Spatial variations in the *magnitude* of **M** were not needed.

Very recently a new class of materials has been found[1] that seems to involve another kind of domain wall: walls across which the *magnitude* changes in space. In these materials magnetic ordering is in competition with the order characterizing superconductivity. They are metallic compounds of magnetic rare earth ions in a superconducting matrix. A proto-

---

H. Suhl is Professor of Physics at the University of California at San Diego.

type is $ErRh_4B_4$. (For a detailed description, particularly of the systematic variation of the magnetic-superconducting properties with substitution of other rare earths, see references 2 and 3.) The effective coupling responsible for the magnetism is so weak that the magnetic order sets in at a temperature T_c well below the superconducting transition temperature T_s (in fact for $ErRh_4B_4$, we have $T_s=8.7$K, $T_c \simeq 1$K). However, once the magnetic order sets in, it tends to diminish or destroy the superconductivity. The reason is that superconductivity (as far as we know to date) requires antiparallel pairing of conduction electron spins. But these spins are in exchange interaction with the magnetic ions (Er in the present case). If they occur only in antiparallel pairs, they cannot acquire a net spin polarization from the aligned magnetic ions, and must sacrifice the polarization energy they would have were they not bound in pairs. When that energy exceeds the pairing energy, the superconductivity tends to be destroyed. It is possible that in some situations a compromise is reached. For example, if the magnetization, instead of aligning uniformly, arranges itself in a spiral order, with several turns of the spiral within the range of the electron-pair wavefunction, the pair can ignore the disruptive effect of the exchange field.[4,5] This possibility aside, there are only two other obvious ways in which coexistence of both phases (in the same volume element) could occur: 1) p-wave superconductivity, in which the pair-wavefunction is antisymmetric and therefore requires parallel-spin pairing, and 2) a ferromagnetic order not mediated by the conduction electrons (the Ruderman-Kittel-Kasuya-Yosida interaction). As for 1), there is so far no experimental evidence for p-wave pairing. As for 2), one could envisage dipolar ferromagnetism (in suitable non-cubic structures), or one could perhaps invoke RKKY coupling by a class of electrons not involved in the superconductivity.

If the possibility 2) occurred in $ErRh_4B_4$, it would be difficult to explain why the superconductivity would cease below T_c, since the relevant electrons no longer see the magnetism. The only pair-breaking effect they would see would be due to coupling of their orbits to any magnetic fields from the aligned ions. Although I doubt that this can be the dominant mechanism, the possibility is analyzed toward the end of this paper. I begin by analyzing the more likely situation: that of significant exchange coupling of conduction electrons to Erbium spins, and for the purposes of this paper I treat the resulting RKKY interaction as the only source of magnetic ordering. Figure 1 shows a schematic plot of the magnetic free energy F_m when the superconductivity is ignored, and the superconducting free energy F_s when the magnetism is ignored, for the case $T_c^0 < T_s^0$ and $F_m(T=0) < F_s(T=0)$. If there were no coupling, the two phases would coexist and the total free energy would be $F_s + F_m$. When a coupling of

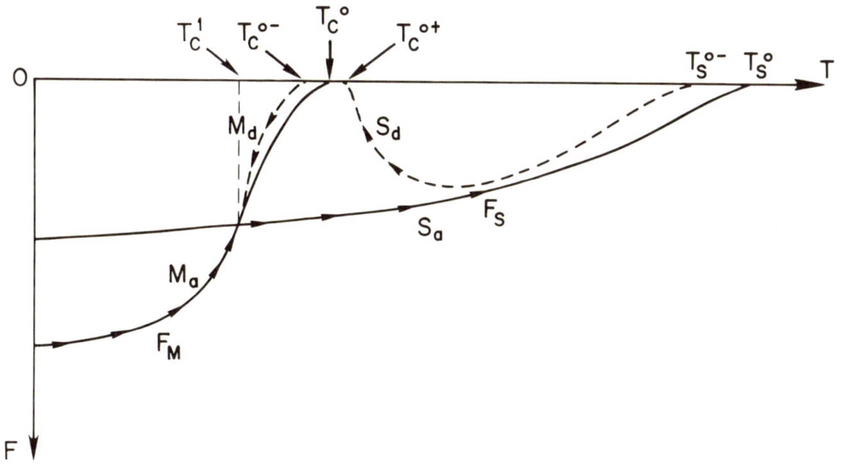

FIG. 1. FREE ENERGY VERSUS ASCENDING AND DESCENDING TEMPERATURE, disregarding the possibility of spiral magnetic structure. In the ascending case a first order transition should occur at T_c^1. In the descending case, the transition should be second order near T_c^0. (*S*—superconductivity, *M*—magnetic, *a*—ascending, *d*—descending.)

sufficient strength is turned on, we see from the figure that an interesting hysteretic situation should arise:

1. Suppose the temperature is lowered from $T > T_s^0$. The sequence of events is as follows:
   a) At a temperature T_s^{0-} somewhat below T_s^0, the sample goes superconducting. T_s is diminished from T_s^0 by the electron coupling to the paramagnetic fluctuations of the magnetization.[6]
   b) Slightly above T_c^0, at T_c^{0+}, the spin fluctuations become critically large, and diminish the superconducting order parameter continuously to zero. Slightly below T_c^0, at T_c^{0-}, a small net magnetization begins to build up from zero. Thus we have a second order transition. (In the very narrow range $T_c^{0-} < T < T_c^{0+}$ the situation is unclear and requires a full renormalization treatment of the transition.)
   c) From T_c^{0-} downwards, there is no more superconductivity (only, perhaps, a fluctuation enhancement of the normal conductivity).
2. Suppose the temperature is raised from $T = 0$.
   a) The sample stays magnetic up to the cross-over at T_c^1 (figure 1). At this point a first order transition should occur. T_c^1 is near neither "natural" transition temperature T_c^0 nor T_s^0. Therefore there are no critical fluctuations in either order parameter capable of reducing the other continuously to zero.

b) As with any first order transition, superheating effects become possible, provided the surface energy between superconducting and magnetic regions is positive.
c) If the superheating can be maintained up to $\sim T_c^{0-}$, a second order transition to the non-magnetic, superconducting state occurs.

This picture is further complicated if so called crypto-magnetic states are admitted. Then as T_c^{0+} is approached from above, spiral ordering could set in without destruction of superconductivity. This transition is weakly first order (the superconducting order parameter undergoes a small discontinuous reduction). Detailed calculations have not yet been performed concerning the further progress as T is diminished. Quite possibly a first order transition to the fully aligned magnetic state eventually occurs.

II. THE FREE ENERGY DENSITY

In this paper I discuss the domain wall energy between magnetic and superconducting regions at temperatures near T_c^1, using the Ginsburg Landau equations for the coupled magnetic and superconducting order parameters M and ψ respectively. Strictly speaking, this is valid only near transitions at which both M and ψ are small, which is not necessarily the case near T_c^1, but the results of such a theory usually are qualitatively valid beyond this limit. We take the Ginsburg-Landau free energy density in the form

$$F = F_S + F_M + F_{SM}$$

where

$$F_S = \frac{\hbar^2}{2m}|\nabla\psi|^2 - \alpha|\psi|^2 + \frac{1}{2}\beta|\psi|^4$$

$$F_M = \frac{1}{2}d^2(\nabla M)^2 - aM^2 + \frac{1}{2}bM^4$$

$$F_{SC} = \frac{1}{2}pM^2|\psi|^2 \quad . \tag{1}$$

Here p is a constant that is positive to ensure that the tendencies to the two types of ordering are in conflict. b and β are positive and almost constant. Also, $a = a_o(T_c^0 - T)$, $\alpha = \alpha_o(T_s^0 - T)$ where a_o and α_o are constants. The use of $\hbar^2/2m$ in F_s is a mere convention; the relevant physical parameter is the superconducting coherence length ξ given by

$$\xi^2 = \hbar^2/(2m\alpha) \quad .$$

d is related to the magnetic coherence length brought about by exchange stiffness. In terms of a and d the magnetic coherence length is l, where

$$l^2 = d^2/a \quad .$$

The energies F_s and F_m separately have minima at $|\psi|^2 = \alpha/\beta$ and at $M^2 = a/b$, both independent of position, and the corresponding energies are $-\alpha^2/2\beta$ and $-a^2/2b$ respectively. In the coupled situation it is therefore useful to introduce reduced variables f and μ according to

$$\psi = \sqrt{\frac{\alpha}{\beta}}\, f$$

$$M = \sqrt{\frac{a}{b}}\, \mu$$

so that the total free energy becomes

$$F = \frac{\alpha^2}{\beta}\left(\xi^2(\nabla f)^2 - f^2 + \frac{1}{2}f^4\right) +$$

$$\frac{a^2}{b}\left(l^2(\nabla\mu)^2 - \mu^2 + \frac{1}{2}\mu^4\right) + \frac{1}{2}\frac{pa\alpha}{b\beta}\mu^2 f^2 \quad .$$

The forms of f and μ that make $\int F dv$ stationary satisfy

$$\xi^2\nabla^2 f + f[1 - (f^2 + \gamma_1\mu^2)] = 0$$

$$l^2\nabla^2\mu + \mu[1 - (\mu^2 + \gamma_2 f^2)] = 0 \qquad (2)$$

where $\gamma_1 = pa/2\alpha b$ and $\gamma_2 = pa/2a\beta$. Now at T_c^1, which is defined by $(F_S)_{\min} = (F_M)_{\min}$, we evidently have $\alpha^2/\beta = a^2/b$, so that at T_c^1, γ_1 and γ_2 become equal, and their common value is

$$\gamma = p/(2\sqrt{\beta b}) \quad .$$

Since we are interested in the domain wall energy only in the vicinity of T_c^1, we shall assume this common value.

Also, we wish specifically to exclude the possibility of coexistence of uniform nonzero values of both μ and f. The uniform solutions of

$$\xi^2\nabla^2 f + f[1 - (f^2 + \gamma\mu^2)] = 0$$

$$l^2\nabla^2\mu + \mu[1 - (\mu^2 + \gamma f^2)] = 0 \tag{3}$$

are the usual ones ($f=0$, $\mu=1$) and ($f=1$, $\mu=0$) and in addition, the "coexistence" solutions

$$f^2 + \gamma\mu^2 = 1$$

$$\mu^2 + \gamma f^2 = 1$$

which give $\mu^2 = f^2 = 1/(1+\gamma)$. We require that this last one be a maximum or a saddle point of F rather than a minimum.

Writing $F = \frac{a^2}{2b}\tilde{F}$, it turns out that at that point $\tilde{F}^2_{\mu f} - \tilde{F}_{\mu\mu}\tilde{F}_{ff} = 16\frac{\gamma-1}{\gamma+1}$, so that it is indeed a saddle point for $\gamma > 1$, whereas ($f=0$, $\mu=1$) and ($f=1$, $\mu=0$) are both minima. When $\gamma < 1$, since $\frac{\partial^2\tilde{F}}{\partial f^2} = -2 + 6f^2 + 2\gamma\mu^2 = \frac{4}{1+\gamma}$ at $f^2 = \mu^2 = \frac{1}{1+\gamma}$, that point is a minimum, and the energy there is $\alpha - f^2 + \frac{1}{2}f^4 - \mu^2 + \frac{1}{2}\mu^2 + \gamma\mu^2 f^2 = -\frac{1}{1+\gamma}$. The points ($f=0$, $\mu=1$) and ($f=1$, $\mu=0$) are saddle points when $\gamma < 1$.

We shall here be concerned with the case $\gamma > 1$. The two minima ($f=0$, $\mu=1$) and ($f=1$, $\mu=0$) with energies $\tilde{F}_{\min} = -\frac{1}{2}$ are then separated by a barrier energy $-\frac{1}{1+\gamma}$, which is the energy of the saddle point $f^2 = \mu^2 = \frac{1}{1+\gamma}$. Unless it turns out that spatial gradients can lower the energy, there will be a barrier to the growth of one kind of domain at the expense of the other as T_c^1 is traversed. This surface energy can be found by solving Eqs. (3).

[We note here that the form $\sim m^2|\psi|^2$ for the coupling energy is not always adequate. As shown elsewhere,[4] for a discussion of the second order transition as T decreases through T_c^0, it is necessary to take into account the non-local character of ψ. If this is not done and the form $m^2|\psi|^2$ is adopted, the effect of the magnetic fluctuations on ψ becomes temperature independent, as the result of a sum-rule. In the discussion of a first order transition, however, taking this matter into account would only produce a refinement of the conclusions.] To find the surface energy we need a solution of the one-dimensional case of Eq. (3), for $\gamma > 1$, in which f varies from 1 to 0 and μ from 0 to 1 as x varies from $-\infty$ to $+\infty$. We discuss three cases:

a) $l << \xi$
b) $\xi << l$
c) $\gamma >> 1$

Case a). μ adjusts itself to the local value of f, because $l\nabla\mu^2$ is negligible:

$$\mu^2 = 1 - \gamma f^2$$

in the region in which $f^2 < \frac{1}{\gamma}$. Where $f^2 > \frac{1}{\gamma}$ we have $\mu = 0$. Thus we must solve

$$\xi^2 f'' + (1 - \gamma) f [1 - f^2(1 + \gamma)] = 0 \quad , \quad f^2 < \frac{1}{\gamma} \tag{4}$$

and

$$\xi^2 f'' + f(1 - f^2) = 0 \quad , \quad f^2 \geq \frac{1}{\gamma} \tag{5}$$

subject to $f = 1$, at $x = -\infty$ and $f = 0$ at $x = +\infty$.

For $f^2 \geq \frac{1}{\gamma}$ the solution has the form

$$f = \frac{1 - A\, e^{2\sqrt{2}x/\xi}}{1 + A\, e^{2\sqrt{2}x/\xi}}$$

where A is a constant.

For $f^2 < \frac{1}{\gamma}$, we write $f\sqrt{1+\gamma} = \varphi$, $\frac{\xi^2}{\gamma - 1} = \bar{\xi}^2$ in which case Eq. (4) becomes

$$\bar{\xi}^2 \varphi'' - \varphi(1 - \varphi^2) = 0 \quad .$$

The solution subject to $f = 0$ at $x = +\infty$ is

$$f = \frac{2\sqrt{2}Be^{x/\xi}}{(1 + B^2e^{2x/\xi})\sqrt{(\gamma+1)}} .$$

A and B are found from the requirement that these expressions be equal to $\frac{1}{\sqrt{\gamma}}$ at a particular x, say $x=0$. This gives $A = \frac{\sqrt{\gamma-1}}{\sqrt{\gamma+1}}$, and $B=(\sqrt{2\gamma}\pm\sqrt{\gamma-1})/\sqrt{\gamma+1}$. The choice of sign is made as follows: Because of the approximation ($\mu^2 = 1-\gamma f^2, f^2 < \frac{1}{\gamma}$), $\mu^2=0$ elsewhere, a discontinuity of slope arises at $x=0$. The slope of the exact solution of course has no such discontinuity. We chose the sign in the expression for B such that the discontinuity is least obtrusive. For $f^2 \geq \frac{1}{\gamma}$ the slope is negative, equal to $\frac{-\sqrt{2}}{2\gamma\xi}$ $(\gamma-1)$, whereas for $f^2 \leq \frac{1}{\gamma}$ it is $\mp\frac{\sqrt{2}\sqrt{\gamma-1}}{2\xi\gamma(\gamma+1)}$. It follows that the upper case should be chosen

$$B = (\sqrt{2\gamma} + \sqrt{\gamma-1})/\sqrt{\gamma+1}$$

in order that the slope, though discontinuous, be at least negative on both sides of $x=0$.

Case b). By symmetry, the solution can be derived from the solution of case a) by the replacements: $f \leftrightarrow \mu$ and $\xi \leftrightarrow l$, $\bar{\xi} \leftrightarrow \bar{l}$ where $\bar{l} = l/\sqrt{\gamma-1}$.

Case c). This case is the simplest (and quite possibly the most relevant). When $\gamma >> 1$ we evidently have $f(x)=0$ $x \geq 0$, and $\mu(x)=0$ $x \leq 0$. Thus we solve

$$\xi^2 f'' + f(1-f^2) = 0$$

subject to $f=1, x=-\infty$, and $f=0, x=0$. A first integral is

$$\frac{\xi^2}{2}(f')^2 + \frac{1}{2}f^2 - \frac{1}{4}f^4 = \text{const.}$$

and, since $f'=0$ at $x=-\infty$, the constant is $+\frac{1}{4}$. Further integration, subject to $f(-\infty)=1, f(0)=0$ gives

$$f(x) = \frac{1-e^{\sqrt{2}x/\xi}}{1+e^{\sqrt{2}x/\xi}} , \qquad x<0$$

$$= 0 \qquad , \qquad x<0 .$$

In exactly the same way

$$\mu(x) = \frac{1 - e^{-\sqrt{2}x/\xi}}{1 + e^{-\sqrt{2}x/\xi}}, \qquad x > 0$$

$$= 0 \qquad , \qquad x < 0 \quad .$$

The free energy associated with this configuration is (recall that $a^2/2b = \alpha^2/2\beta$ at T_c^1),

$$\frac{a^2}{b} \int \left\{ \left[\frac{\xi^2}{2}(f')^2 - \frac{1}{2}f^2 + \frac{1}{4}f^4 \right] \theta(-x) + \right.$$

$$\left. \left(\frac{l^2}{2}\mu'^2 - \frac{1}{2}\mu^2 + \frac{1}{4}\mu^4 \right) \theta(x) \right\} dx$$

$$= \frac{a^2}{b} \int \left\{ \left[\xi^2(f')^2 + l^2(\mu')^2 \right] - \frac{1}{4} \right\} dx$$

where $\theta(x) = 0$ or 1 according as $x \lessgtr 0$.

On the other hand, the energy of either the uniformly magnetic or uniformly superconducting state is $-a^2/4b\int dx$. Thus it follows that the surface energy is positive, and equal to

$$E_S = \frac{a^2}{b} \left[\xi^2 \int_{-\infty}^{0} (f')^2 dx + l^2 \int_{0}^{\infty} (\mu')^2 dx \right]$$

$$= \frac{a^2}{\sqrt{2}b}(l + \xi) \int_0^\infty \frac{dx}{\cosh^4 x}$$

$$= \frac{2}{3\sqrt{2}} \frac{a^2}{b}(l + \xi) \quad .$$

From this we can estimate the temperature width of the hysteretic region. Suppose one is going up in temperature from $T = T_c^1$ to $T = T_c^1 + \Delta T$. The cost of maintaining a volume $\frac{4}{3}\pi r^3$ of the sample in the magnetic state is

$$-\frac{4}{3}\pi r^3 \left[\frac{a_0^2\,(T_c^1 + \Delta T - T_c^0)^2}{2b} - \frac{\alpha_0^2\,(T_c^1 - T_s)^2}{2\beta}\right]$$

$$= \frac{4}{3}\pi r^3 \left[\frac{a_0^2\,(T_c^0 - T_c^1)\Delta T}{b}\right] \quad .$$

On the other hand the surface energy is $4\pi r^2 E_S$. The two are equal when

$$\frac{r}{3}\Delta T = (l + \xi)\,(T_c^0 - T_c^1) \quad .$$

But superconducting nuclei cannot possibly be smaller than $r \sim \xi$; therefore, the maximum hysteresis width on the way up is

$$\Delta T = 3\left(1 + \frac{l}{\xi}\right)(T_c^0 - T_c^1) \quad .$$

Similarly on the way down it is

$$\Delta T' = 3\left(1 + \frac{\xi}{l}\right)(T_c^0 - T_c^1) \quad .$$

III. DIAMAGNETIC EFFECTS

Maekawa and Tachiki[7] have made a detailed study of the diamagnetic aspects of these materials when they are in the paramagnetic regime, i.e., above T_c^0. It is of some interest to consider these effects at and below T_c^0 also, even though most likely the exchange coupling of the superconducting electrons to the localized spins dominates the character of the transition. We consider here only an extreme situation, one in which the ordering mechanism of the rare earth spins does *not at all* involve the superconducting electrons. For example, for low enough symmetry of the magnetic sites, the ferromagnetism might be of dipolar origin, or conceivably, it may arise by indirect interaction via conduction electrons belonging to a band "orthogonal" to that of the superconducting electrons. Here we consider only this latter case.

The total free energy now takes the form

$$F = F_S + F_M + H^2/8\pi - \mathbf{H} \cdot \mathbf{M}$$

where F_M has the same form as before and

$$F = \frac{1}{2m} \left| \left(-i\hbar \nabla - \frac{2e\mathbf{A}}{c}\right)\psi \right|^2 - \alpha|\psi|^2 + \frac{1}{2}\beta|\psi|^4 \quad .$$

F_{SM}, by hypothesis, is absent, and the coupling of the two order parameters now takes place through the electromagnetic field only.

The resulting variational equations are

$$\frac{1}{2m}\left(-i\hbar\nabla - \frac{2e\mathbf{A}}{c}\right)^2 \psi - \alpha\psi + \beta|\psi|^2\psi = 0 \tag{6a}$$

$$-d^2\nabla^2\mathbf{M} - a\mathbf{M} + bM^2\,\mathbf{M} - \mathbf{H} = 0 \tag{6b}$$

$$\nabla\times\nabla\times\mathbf{A} = -\frac{4\pi e\hbar}{imc}(\psi^*\nabla\psi - \psi\nabla\psi^*) - \frac{16\pi e^2}{mc^2}|\psi|^2\mathbf{A} + 4\pi\nabla\times\mathbf{M} \quad . \tag{6c}$$

For the one-dimensional variation considered here, with $\mathbf{M} = (0, 0, M(x))$, $\mathbf{A} = (0, A(x), 0)$, $\mathbf{H} = (0, 0, H(x))$, $\psi = fe^{i\theta}$, $\theta =$ constant, these equations reduce to

$$-\frac{h^2}{2m}\frac{\partial^2 f}{\partial x^2} - \alpha f + \beta f^3 + \frac{2e^2}{mc^2}A^2 f = 0$$

$$-d^2\frac{\partial^2 M}{\partial x^2} - aM + bM^3 - H = 0$$

$$-\frac{\partial^2 A}{\partial x^2} + \frac{A}{\lambda^2} + 4\pi\frac{\partial M}{\partial x} = 0 \tag{7}$$

$$\frac{\partial H}{\partial x} = \frac{A}{\lambda^2}$$

$$\lambda^2 = mc^2\beta/(16\pi e^2\alpha) \quad .$$

Note that Eqs. (6) at once give the results of Maekawa and Tachiki for the paramagnetic case $a = -|a|$, at least for the case of very small d. Then $\mathbf{M} = \mathbf{H}/|a| = \chi\mathbf{H}$ where χ is the paramagnetic susceptibility. Since $\nabla \times \mathbf{H} = 4\pi\mathbf{J}/c$, the last equation of (6) then reads $\nabla \times \nabla \times \mathbf{A} = \frac{4\pi}{c}\mathbf{j}(1 + 4\pi\chi)$. For the case of a constant phase of ψ, this may be written $\nabla \times \nabla \times \mathbf{A} = -\mathbf{A}/\lambda^2$, where

$$\lambda_{\text{eff}}^2 = \lambda^2/(1 + 4\pi\chi)$$

so that, as T_c^0 is approached from above, and χ becomes large, the effective penetration length shortens, and the superconductor tends to have type I properties. Sufficiently below T_c^0, the same conclusion applies. Sufficiently below T_c^0, it is possible to write $M = M_0 + \chi_F H$, with $M_0 = \text{const}$. For very small l, the second of Eqs. (6) gives

$$\chi_F\,(3bM_0^2 - a) = 1$$

where

$$\chi_F = 1/2a$$

and now $\lambda_{\text{eff}}^2 = A^3/(1 + \chi_F)$. Thus below T_c^0, also, there is some tendency towards type I behavior. Very close to T_c^0, the variation of M with H is non-analytic. To obtain some idea of the behavior there, we consider the case $T = T_c^0$, neglecting as before the $d^2\nabla^2\mathbf{M}$ term.

Then $a = 0$, and $M = (H/b)^{1/3}$ and so we have

$$-\frac{\partial^2 A}{\partial x^2} = -A/\lambda^2 - 4\pi\frac{\partial}{\partial x}\left(\frac{H}{b}\right)^{1/3}$$

together with

$$\frac{\partial H}{\partial x} = A/\lambda^2 \quad .$$

It follows that

$$\frac{\partial^2 H}{\partial x^2} = (H + 4\pi(H/b)^{1/3})/\lambda^2 \tag{8}$$

if we suppose that $H'' = 0$ when $H = 0$. A first integral of (8) is

$$\left(\frac{\partial H}{\partial x}\right)^2 = (H^2 + (6\pi/b)^{1/3} H^{4/3})/\lambda^2$$

if $H' = 0$ when $H = 0$. We seek a solution that has $H = 0$ at $x = \infty$ and $H = H_0$, the applied tangential field at the sample boundary. The substitution $H = y^3$ turns this equation into

$$\int \frac{dy}{\sqrt{y^2 + \left(\frac{6\pi}{b}\right)^{1/3}}} = (-x + \text{const.})/3\lambda$$

(the minus sign being chosen to obtain decay in the $+x$ direction). The results are $y = (6\pi/b)^{1/3}$ [$\sinh(\Lambda - x)/3\lambda$], whence $H = (6\pi/b)\sinh^3[(\Lambda - x)/3\lambda]$. The constant Λ is found from $H_0 = (6\pi/b)\sinh^3(\Lambda/3\lambda)$. This solution persists up to $x = \Lambda$. For $x > \Lambda$, the solution must be $H = 0$. The derivatives $(\partial^n H/\partial x^n)_{x=\Lambda+0}$ and $(\partial^n H/\partial x^n)_{x=\Lambda-0}$ are continuous up to and including $n = 2$, so that no difficulties arise with the solution (8). Hence at $T = T_c^0$ we have a field-dependent penetration depth, equal to

$$\Lambda = 3\lambda \sinh^{-1}\left(\frac{bH_0}{6\pi}\right)^{1/3} ,$$

which tends to zero with H_0. This suggests that at $T = T_c^0$, the behavior is type I. Thus there will always be a certain neighborhood around T_c^0 in which type I behavior may be expected. Accordingly, the surface energy will be positive, and the transition may be expected to be first order and hysteretic.

IV. STATES WITH SPIRAL MAGNETIZATION

In a previous publication, I discussed states with uniform ψ and spiral M that were solutions of the Ginsburg Landau equations when only the exchange mechanism is active. In this section we note that such a solution also exists when only the indirect diamagnetic effects of the magnetization are considered.[8] To see this, we write $\psi = \text{const.} = f$, $\mathbf{M} = (M_0 \cos qz, M_0 \sin qz, 0)$. $\mathbf{H} = (h \cos qz, h \sin qz, 0)$, $\mathbf{A} = -\lambda^2 \nabla \times \mathbf{H} = -\lambda^2 hq (\cos qz, \sin qz, 0) = -\lambda^2 q\mathbf{H}$ in Eqs. (6), with the result

$$\alpha = \beta f^2 + \frac{\lambda^2}{8f^2} q^2 h^2 \qquad \text{(from 6a)} \qquad (9a)$$

$$bM_0^2 = a - d^2q^2 + \frac{h}{M_0} \qquad \text{(from 6b)} \qquad (9b)$$

$$h = -4\pi M_0/(1 + q^2\lambda) \qquad \text{(from 6c)} \quad . \qquad (9c)$$

The usual solution is $q=0$ and gives $f^2=\alpha/\beta$, $M_0^2=(a-4\pi/b)$ and $h=-4\pi M_0$ corresponding to complete expulsion of the flux B. Note that the effective T_c^0 is depressed as the result of the expulsion: $k_BT_c^0 \rightarrow k_BT_c^0 - 4\pi/a_o$; in whichever direction the magnetization attempts to form, the field due to the flux expulsion attempts to reduce it. For finite q this effect will be diminished provided

$$d^2q^2 + \frac{4\pi}{1+q^2\lambda^2} < 4\pi$$

i.e., provided

$$q^2\lambda^2 < \frac{4\pi\lambda^2}{d^2} - 1 \quad .$$

This can always be arranged provided $\lambda > d/2\sqrt{\pi}$, which is normally satisfied. The flux expulsion is no longer total, a condition that favors the establishment of the full magnetization. To see what a state with $q \neq 0$ implies for the gap parameter f, we note that

$$\lambda^2 = \frac{mc^2}{16\pi e^2 f^2} \quad , \quad m = \frac{2eh\varkappa}{c}\left(\frac{2\pi}{\beta}\right)^{1/2}$$

where $\varkappa = \lambda/\xi$. Then the equation for f becomes (with ϕ_0 the quantum of flux),

$$\alpha = \beta f^2 + \frac{\phi_0\varkappa}{4\pi}\left(\frac{2\pi}{\beta}\right)^{1/2} \frac{q^2M_0^2}{\left(f^2 + \frac{\phi_0\varkappa q^2}{\lambda\pi^2}\left(\frac{2\pi}{\beta}\right)^{1/2}\right)^2}$$

This equation has a solution (evolving continuously from $f^2=\alpha/\beta$ as q^2 is increased from zero) provided that

$$\alpha > 2^{-2/3}\left(\frac{\phi_0 \varkappa}{4\pi}\left(\frac{2\pi}{\beta}\right)^{1/2} q^2 M_0^2\right)^{1/3} \beta^{2/3} - \frac{\phi_0 \varkappa q^2}{8\pi^2}\left(\frac{2\pi}{\beta}\right)^{1/2} .$$

(Strictly speaking, the effect on M_0 of the change in f according to Eq. (9b) and (9c) should also be considered, but is presumably small and is neglected here.) The value of f^2 is largest at $q=0$; therefore a state with $q \neq 0$ involves some sacrifice of superconducting energy. This loss may well be more than compensated by a lowering of magnetic energy for $q \neq 0$. However, at the present stage of experimental work it seems premature to follow this speculative line any further.

ACKNOWLEDGMENTS

The author acknowledges helpful discussions with Dr. M. Brian Maple and Mr. B. MacKay.

REFERENCES

1. B. T. Matthias, E. Corenzwit, J. M. Vandenberg, and H. E. Barz, Proc. Natl. Acad. Sci. U.S. **74**, 1334 (1977).
2. W. A. Fertig, D. C. Johnston, L. E. DeLong, R. W. McCallum, M. B. Maple, and B. T. Matthias, Phys. Rev. Lett. **38**, 987 (1977).
3. M. B. Maple, C. Hamaker, D. C. Johnston, H. B. MacKay, and L. D. Woolf, to be published in J. Less-Common Met.
4. H. Suhl, J. Less-Common Met. **62**, 225 (1978).
5. P. W. Anderson and H. Suhl, Phys. Rev. **116**, 898 (1959).
6. A. A. Abrikosov and L. P. Gor'kov, Sov. Phys.—JETP **12**, 1243 (1961).
7. S. Maekawa and M. Tachiki, preprint.
8. As this work was completed, I learned of an independent derivation of a periodic state from microscopic theory by Matsumoto, Tachiki, and Umezawa (preprint), and another one by Blount and Varma closely paralleling the derivation given here.

NUCLEAR HYDRODYNAMICS IN A RELATIVISTIC MEAN FIELD THEORY

by J. D. Walecka

ABSTRACT

We review a model relativistic quantum field theory of the nuclear many-body problem containing baryons and vector and scalar mesons. We also review the properties of nuclear and neutron matter at all temperatures and densities and of finite nuclei computed in the mean-field approximation. The justification for this approximation is discussed. The model is then used as a basis for discussing the *hydrodynamic flow* of nuclear matter. This is of interest in heavy-ion reactions, in nuclear spectroscopy, and in astrophysics. The mean-field equations obtained by minimizing the energy functional at fixed baryon and momentum densities are derived and used to study the hydrodynamic mass at all densities and flow velocities. The limiting cases of non-relativistic and extreme-relativistic motion are examined in detail. The model is explicitly covariant, and the results are shown to be equivalent to those obtained with appropriate Lorentz transformations.

I. INTRODUCTION

In 1933 Felix Bloch used the linearized equations for hydrodynamic flow to describe the electron cloud in the atom and to carry out his calculation of the energy loss of a charged particle in matter.[1] His elegant treatment of the collective dipole excitations about the mean-field ground state is a major contribution to the quantum theory of many-particle systems. Bloch's approach can be extended to describe collective charge oscillations of any multipolarity in atoms,[2] and also collective spin-waves.[3] If the

J. D. Walecka is Professor of Physics at Stanford University.

electron cloud in an atom is divided into a set of core electrons treated collectively with Bloch's approach and a set of loosely bound valence electrons, then a reasonable picture is obtained of real neutral atoms including binding energies, charge densities, and photoionization cross sections.[4]

There are many reasons why it is desirable also to have a model for the hydrodynamic flow of *nuclear* matter. The study of collective excitations of nuclei about the mean-field ground state is central to nuclear physics. Furthermore, heavy-ion reactions where large aggregates of nuclear matter are fired into contact offer the possibility of directly observing nuclear hydrodynamics. In addition, the hydrodynamics of nuclear matter forms an essential ingredient in astrophysics in the discussion of supernovae and the formation of neutron stars.

Consider first the case of a *static* neutron star, which is nothing more than an enormous nucleus held together by gravitational attraction. The interior density in a neutron star can be a factor of 10-20 higher than in the interior of observed terrestrial nuclei. To discuss the properties of cold, condensed stellar objects such as neutron stars, it is necessary to know the stress tensor $T_{\mu\nu}$, the source in Einstein's field equations, from nuclear matter densities upwards. For a uniform fluid at rest, which locally describes the situation in a static neutron star, the stress tensor takes the form

$$T_{\mu\nu} = P\delta_{\mu\nu} + (\varrho + P/c^2)u_\mu u_\nu \tag{1.1}$$

where P is the pressure, ϱ the mass density, and $u_\mu = (\mathbf{o}, ic)$ is the four velocity of the fluid. What is thus required at the outset is $P(\varrho)$, or the equation of state of nuclear matter at high density.

Any description of nuclear matter under extreme conditions of density, flow, and temperature requires a more general starting point than the usual theory of nuclear structure based on non-relativistic nucleons interacting through static two-body potentials. There are several reasons for this. For example:

1) At nuclear densities, the velocity is already $v/c \sim 1/4$ for the fastest nucleons at the Fermi surface. As the density is raised the velocities increase, and the relativistic propagation of the baryons, and relativistic propagation and retardation of the mesons generating the interactions, must be taken into account.

2) It is an approximation to describe nuclei in terms of only nucleon coordinates, and a proper description, particularly at high densities where real mesons may be present, must explicitly take into account the meson degrees of freedom.

3) Correct, causal restrictions on the propagation of the modes of excitation of the interacting system must be contained in the theory.

The only consistent theoretical framework for dealing with this nuclear system that I am aware of is a relativistic quantum field theory. Given an appropriate Lagrangian density, we in principle have a complete calculational framework with which to answer any physical question that can be asked about the system. With a relativistic theory that explicitly includes the meson degrees of freedom, we can in principle make reliable extrapolations away from the narrow window on the equation of state of nuclear matter provided by observed nuclei. We must, of course, work with a renormalizable quantum field theory in order to obtain finite results for physical quantities.

The purpose of the present paper is to calculate the properties of hydrodynamic flow of cold nuclear matter within the framework of a model relativistic quantum field theory under study at Stanford for the past few years.[5–14] Many of the ideas in the model are similar in spirit to those of Johnson and Teller[15] as developed by Duerr,[16] and important recent applications of the present model and similar ones[17,18] to the nuclear surface and properties of finite nuclei have also been made by Kerman and Miller,[19] Boguta and Bodmer,[20] and Boguta and Rafelski.[21]

II. THE MODEL

As a model, consider the fields in table 1. We assume that the neutral scalar meson couples to the scalar baryon density through a coupling $g_s\bar{\psi}\psi\phi$ and that the vector meson couples to the conserved baryon current through $ig_v\bar{\psi}\gamma_\lambda\psi V_\lambda$.

TABLE 1

FIELDS IN THE MODEL

Field	Description	(Particles)	Mass × $(c/\hbar)$
ψ	baryon	$(p,n, \ldots)$	M
ϕ	neutral scalar meson	(σ)	m_s
V_λ	neutral vector meson	(ω)	m_v

As motivation for this model, we observe that in the limit of heavy, static baryons the one-meson exchange graphs give rise to an effective nucleon-nucleon potential of the form

$$V_{\text{eff}} = \frac{g_v^2}{4\pi}\frac{e^{-m_v r}}{r} - \frac{g_s^2}{4\pi c^2}\frac{e^{-m_s r}}{r} \tag{2.1}$$

and for appropriate choices of the coupling constants and masses, this potential can be made attractive at large separations and repulsive at short distances in accord with the observed nucleon-nucleon force. The reader will ask, "where are the pions?" In fact, the effects of the pion interaction with baryons largely average out in the description of the bulk properties of nuclear matter because of the strong spin dependence of the coupling. Eventually we must include additional fields for (**π**, **ρ**, etc.) if we hope to achieve a truly quantitative description of the nuclear system.

The field equations for this model are

$$(\Box - m_s^2)\phi = -(g_s/c^2)\bar{\psi}\psi \tag{2.2}$$

$$\frac{\partial}{\partial x_\nu}F_{\mu\nu} + m_v^2 V_\mu = ig_v\bar{\psi}\gamma_\mu\psi \tag{2.3}$$

$$[\gamma_\mu(\frac{\partial}{\partial x_\mu} - \frac{ig_v}{\hbar c}V_\mu) + (M - \frac{g_s}{\hbar c}\phi)]\psi = 0 \quad . \tag{2.4}$$

Here $F_{\mu\nu} \equiv \partial V_\nu/\partial x_\mu - \partial V_\mu/\partial x_\nu$ is the antisymmetric field tensor. Equation (2.2) is simply the Klein-Gordon equation with a scalar source; Eq. (2.3) looks like massive QED with the conserved baryon current

$$B_\mu = i\bar{\psi}\gamma_\mu\psi \tag{2.5}$$

rather than the (conserved) electromagnetic current as source; and Eq. (2.4) is the Dirac equation with the scalar and vector fields introduced in a "minimal" fashion. The free field Lagrangians for this system are contained in any text, and the interaction Lagrangian density is just

$$\mathcal{L}' = ig_v\bar{\psi}\gamma_\mu\psi V_\mu + g_s\bar{\psi}\psi\phi \quad . \tag{2.6}$$

Knowing the Lagrangian density, we can construct the stress tensor $T_{\mu\nu}$ in the canonical fashion, and from this the four-momentum operators for the system (**P**, iH/c); from the expectation value of $T_{\mu\nu}$ for a uniform system, we have the equation of state through Eq. (1.1).

The theory is Lorentz covariant, and the procedure for quantizing to get the quantum field theory is well known. The theory is renormalizable, for it is similar to massive QED with a conserved current and an additional scalar interaction. Since the dimensionless coupling constants $C_v^2 = (g_v^2/\hbar c)(M^2/m_v^2)$ and $C_s^2 = (g_s^2/\hbar c^3)(M^2/m_s^2)$ are large, we have a strong-coupling theory and have not really made much progress unless a sensible starting solution to the theory can be found. The main point of the present work is that we can find a simple solution to the field equations that *becomes increasingly valid as the density of the system increases*, and which thus provides a meaningful starting point for computing modifications of the equation of state through the relativistic quantum field theory and standard many-body techniques[22] in a consistent fashion.

Consider a uniform system of baryons in a box of volume Ω. As the baryon density increases, so do the source terms on the right-hand sides of Eqs. (2.2, 2.3). When the sources are large, the meson fields can be replaced by their expectation values, which are then classical fields.

$$\phi \rightarrow <\phi> \equiv \phi_0 \tag{2.7}$$

$$V_\lambda \rightarrow <V_\lambda> \equiv i\delta_{\lambda 4} V_0 \tag{2.8}$$

and Eqs. (2.2, 2.3) can be immediately solved for a static, uniform system to give

$$\phi_0 = \frac{g_s}{m_s^2 c^2} <\bar{\psi}\psi> \equiv \frac{g_s}{m_s^2 c^2} \varrho_s \tag{2.9}$$

$$V_0 = \frac{g_v}{m_v^2} <\psi^\dagger\psi> \equiv \frac{g_v}{m_v^2} \varrho_B \quad . \tag{2.10}$$

When these classical fields are substituted into the Dirac equation (2.4), this equation is linearized and may be solved exactly. The stationary state solutions for a uniform system $\psi = \exp(-i\epsilon ct + i\mathbf{k}\cdot\mathbf{x})\, U(\mathbf{k},\lambda)$ satisfy a modified Dirac equation with effective mass

$$M^* = M - (g_s/\hbar c)\phi_0 \tag{2.11}$$

and energy spectrum

$$\epsilon_k^{(\pm)} = (g_v/\hbar c)V_0 \pm (\mathbf{k}^2 + M^{*2})^{1/2} \quad . \tag{2.12}$$

The solutions satisfy the relation

$$\bar{U}U = \frac{M^*}{(M^{*2} + \mathbf{k}^2)^{1/2}} U^\dagger U \quad . \tag{2.13}$$

The scalar density is reduced relative to the baryon density by Lorentz contraction. The Dirac fields may be expanded in terms of the solutions to the

Dirac equation with coefficients $A_{k\lambda}$, $B_{k\lambda}$ in standard fashion, and the Hamiltonian density becomes

$$\hat{\mathcal{H}}_{MFT} = g_v V_0 \varrho_B +$$

$$(\hbar c/\Omega)\Sigma_{k\lambda}(\mathbf{k}^2 + M^{*2})^{1/2}(A^{\dagger}_{k\lambda}A_{k\lambda} + B^{\dagger}_{k\lambda}B_{k\lambda})$$

$$+ (m_s^2c^2/2)\phi_0^2 - (m_v^2/2)V_0^2 \quad . \tag{2.14}$$

The new normal-mode amplitudes can be interpreted as creation and destruction operators and the system quantized in the canonical manner. Here we have chosen to define the *mean field theory* (MFT) by normal ordering in these creation and destruction operators. Since the Hamiltonian density in Eq. (2.14) is diagonal, the exact solution for the mean field theory is known and all properties of the system can be computed within this exactly solvable model.

Consider uniform nuclear matter. The ground state of Eq. (2.14) is obtained by filling the states with wave number l and spin-isospin degeneracy γ up to the Fermi level l_F. Here we choose to measure all lengths in units of $1/M \equiv \hbar/m_pc = .2103F$ and all energies in units of $m_bc^2 \equiv m_pc^2 = 938.3$ MeV. The equation of state computed from Eqs. (2.14, 1.1) in the MFT is

$$e = \frac{1}{2}C_v^2n_B^2 + \frac{1}{2C_s^2}(1-\chi)^2 +$$

$$\frac{\gamma}{(2\pi)^3}\int_0^{l_F} d\mathbf{t}(t^2+\chi^2)^{1/2} \tag{2.15}$$

$$p = \frac{1}{2}C_v^2n_B^2 - \frac{1}{2C_s^2}(1-\chi)^2 +$$

$$\frac{1}{3}\frac{\gamma}{(2\pi)^3}\int_0^{l_F} d\mathbf{t}\frac{t^2}{(t^2+\chi^2)^{1/2}} \tag{2.16}$$

$$n_B = \frac{\gamma}{(2\pi)^3}\int_0^{l_F} d\mathbf{t} \tag{2.17}$$

where e, p, and n_B are the dimensionless energy density, pressure, and baryon density. Here

$$\chi \equiv M^*/M \tag{2.18}$$

is the effective mass. The first two terms in Eqs. (2.15, 2.16) arise from the mass terms for the vector and scalar fields. The final two terms are those of a relativistic Fermi gas of baryons of mass χ. The equation of state is given in parametric form $e(n_B)$, $p(n_B)$. It remains to determine the effective mass. This can be done by minimizing $e(\chi)$ with respect to the parameter χ (equivalently, with respect to ϕ_0), which leads to the self-consistency relation

$$\chi = [1 + C_s^2 \frac{\gamma}{(2\pi)^3} \int_0^{t_F} dt \frac{1}{(t^2 + \chi^2)^{1/2}}]^{-1} \quad . \tag{2.19}$$

This is identical to the scalar meson field equation (2.9) where relation (2.13) has been used in the computation of the scalar density. It is evident that at high density $t_F \rightarrow \infty$, the effective mass goes to zero and the energy density is dominated by the vector repulsion, as is the pressure. In this limit $p \rightarrow e$, implying that the thermodynamic speed of sound approaches the velocity of light from below. This property of a pure vector interaction at high density was first pointed out by Zel'dovich.[23] At lower densities, the attractive scalar interaction can be made to dominate if the coupling constants are chosen properly. The nuclear medium then *saturates*. The coupling constants can be chosen to reproduce the equilibrium properties of nuclear matter[22]:

$$\{\gamma = 4, (E - Bm_b c^2)/B = -15.75 \,\text{MeV}, k_F = 1.42F^{-1}\} \quad .$$

Accordingly,

$$C_v^2 \equiv (g_v^2/\hbar c)(M^2/m_v^2) = 195.7 \tag{2.20}$$

$$C_s^2 \equiv (g_s^2/\hbar c^3)(M^2/m_s^2) = 266.9 \quad . \tag{2.21}$$

Note that only the ratios of coupling constants to meson masses enter here. The resulting saturation curve is shown in figure 1.[5,6] The corresponding curve for neutron matter obtained by simply replacing $\gamma = 2$ is also shown in figure 1. Many other properties of nuclear matter are now predicted from the two determined constants in Eqs. (2.20, 2.21). These are the only two parameters in the MFT of nuclear matter, and all other properties may now be calculated. The calculated properties are compared with experiment in table 2. The quantity a_4 is the coefficient in the symmetry energy of the semi-empirical mass formula, and K_v is the bulk modulus.[22] If the scalar field is identified as the σ meson (a broad two-pion resonance at $m_s/M \cong .59$) and the vector field with the ω meson ($m_v/M = .835$), then we obtain from Eqs. (2.20, 2.21) the coupling constants in table 2. These are

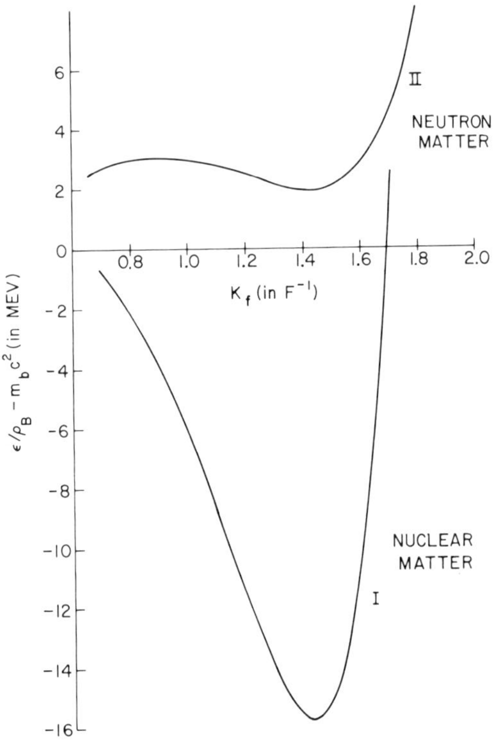

FIG. 1. SATURATION CURVE FOR NUCLEAR MATTER. Coupling constants chosen to fit value and position of minimum. Prediction for neutron matter ($\gamma = 2$) also shown.[6]

TABLE 2

CALCULATED PROPERTIES OF NUCLEAR MATTER AND BASIC COUPLINGS[6]

	a_4	M^*/M	K_v^{-1}	$g_s^2/4\pi\hbar c^3$	$g_v^2/4\pi\hbar c$
Theory	22.0 MeV	0.56	550 MeV	7.39	10.8
Experiment	23.5 MeV	(0.6)	(200-300 MeV)	8.2(σ)	17.3(ω)

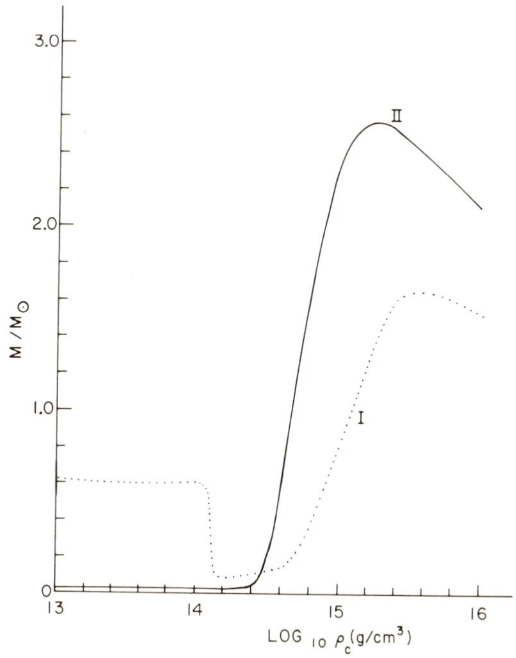

FIG. 2. CALCULATED MASS OF NEUTRON STAR as function of central density-curve II. Curve I comes from integrating the equation of state of Baym et al.[6]

compared with values obtained from a phase-shift analysis of *N-N* scattering.[6]

The equation of state for neutron matter gives the stress tensor through Eq. (1.1). The Tolman-Oppenheimer-Volkoff equations for the metric in general relativity can now be integrated to give the mass of a static neutron star as a function of central density. The result is shown in figure 2.[6] The MFT gives a maximum mass for a neutron star of $M = 2.57\ M_\odot$. There are arguments that the x ray binary pulsar in Vela X1 has a mass $M > 1.4 M_\odot$.

To extend the previous analysis to finite systems with a fixed number of baryons, we allow spatial variations in the mean fields $\phi_0(|\mathbf{x}|)$, $V_0(|\mathbf{x}|)$. This adds the terms $(c^2 m_s^2/2)\,(\nabla\phi_0)^2 - (m_v^2/2)\,(\nabla V_0)^2$ to the Hamiltonian density. If it is assumed that these fields vary slowly enough so that at each point the baryons can be treated as moving in locally constant fields, then the previous analysis from nuclear matter applies. The ground-state of the system is obtained by minimizing the total energy E as a function of the local Fermi wave number $k_F(|\mathbf{x}|)$. The constraint of a fixed number of baryons B can be incorporated through a Lagrange multiplier μ. Thus we set

$$\delta E - \mu\delta B = 0 \quad . \tag{2.22}$$

Since ϕ_0 and V_0 satisfy the field equations

$$(\nabla^2 - m_s^2)\phi_0(r) = -(g_s/c^2)\varrho_s(r) \tag{2.23}$$

$$(\nabla^2 - m_v^2)V_0(r) = -g_v\varrho_B(r) \tag{2.24}$$

the variations with respect to these quantities vanish and they may be held constant during the variation. The result of the variational principle is that $k_F(r)$ must satisfy the equation

$$g_v V_0 + \hbar c\,(k_F^2 + M^{*2})^{1/2} = \mu = \text{const.} \tag{2.25}$$

which can be used to find $k_F(\phi_0, V_0, \mu; r)$. The local baryon and scalar densities are expressed in terms of $k_F(r)$ and $M^*(r)$ as

$$\varrho_B = \frac{\gamma}{(2\pi)^3}\int_0^{k_F(r)} d\mathbf{k} \tag{2.26}$$

$$\varrho_s = \frac{\gamma}{(2\pi)^3}\int_0^{k_F(r)} d\mathbf{k}\,\frac{M^*}{(M^{*2} + k^2)^{1/2}} \quad . \tag{2.27}$$

Since in this Thomas-Fermi MFT the baryon and scalar densities vanish identically past a certain radius r_0, the asymptotic form of the solution is known, and hence $(V_0'/V_0)_{r_0}$ and $(\phi_0'/\phi_0)_{r_0}$ are determined. The coupled, non-linear, differential Eqs. (2.23, 2.24) can be integrated in from r_0, and μ and $V_0(r_0)$ (or $\phi_0(r_0)$) varied until a solution is found that satisfies the interior boundary conditions $V_0'(0) = \phi_0'(0) = 0$. The total number of baryons is determined at the end of the calculation by integrating Eq. (2.26). This has been carried out by Serr and Walecka.[13] There are now two additional parameters in the theory, since m_s and m_v must be specified independently. In Serr and Walecka[13] it is assumed that the properties of nuclear matter are described as in the previous discussion and that Eqs. (2.20, 2.21) hold; m_s/M is then chosen to fit the observed surface thickness of nuclei $t \cong 2.4F$. The dependence on m_v/m_s is not strong, and a value $m_v/m_s = 1.5$ is chosen in rough accord with the observed meson masses ($m_v/m_s \cong 1.42$). Calculated nuclear properties for the system with $B = 40$ are shown in table 3.[13] (Here the Coulomb interaction is neglected so that in equilibrium $N = Z = B/2$). a_2 is the coefficient in the surface energy in the semi-empirical mass formula[22] and is obtained by fitting the calculated energy to an expression $E - (m_b c^2)B = (-15.75\text{MeV})B + a_2 B^{2/3}$. R_0 is the parameter in the half-

TABLE 3

CALCULATED QUANTITIES FOR B = 40.[13] Here $N = Z$.

	a_2/t	R_0	α_{max}	m_s/M
Theory	13.1 MeV/F	1.03F	1.91 MeV	.518
Experiment	7.4 MeV/F	1.07F	1.80 MeV*	.59

*Obtained from $<\alpha(r)>$ in Sc^{41}.

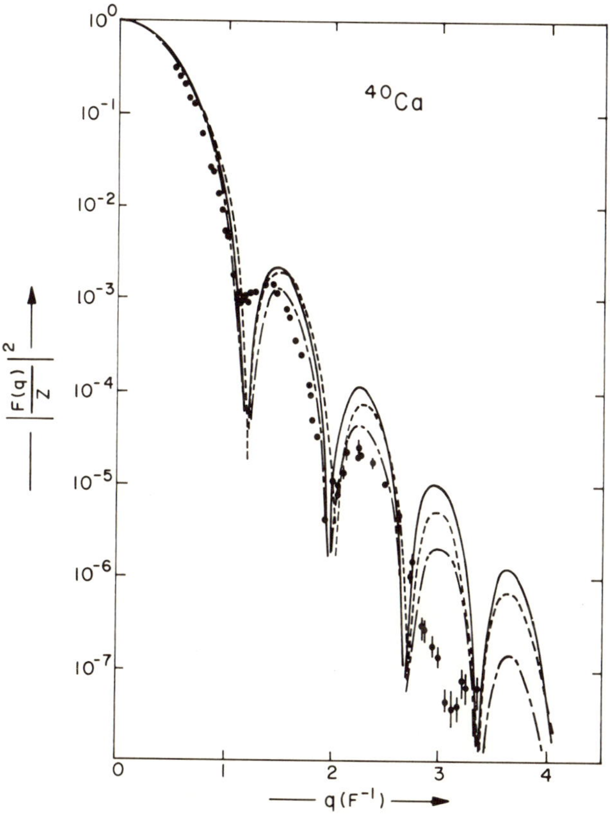

FIG. 3. CALCULATED AND EXPERIMENTAL FORM FACTOR FOR $_{20}Ca^{40}$. Dot-dashed curve includes single-nucleon form factor. Dashed curve includes vacuum fluctuations.[13]

density radius $R_{1/2} = R_0 B^{1/3}$. To test the quality of the calculated charge density, the Fourier transform of the calculated density is compared with the experimental values of the elastic electron-scattering form factor $[d\sigma/d\Omega]_{el}/[d\sigma/d\Omega]_{point}$ for ${}_{20}Ca^{40}$ in figure 3. Also shown is the result of folding in the internal electromagnetic structure of the nucleons through the experimental single-nucleon form factor. The model evidently provides a decent description of the nuclear form factor except for the highest momentum transfers, which probe short-distance spatial variations of the charge density.

The Foldy-Wouthuysen reduction of the single-particle Dirac equation for non-relativistic nucleons moving in the potential generated by the scalar and vector fields $\phi_0(r)$ and $V_0(r)$ allows us to identify the effective single-particle *spin-orbit interaction*

$$V_{so}(r) = \frac{1}{2M^2 r}\left(g_v \frac{dV_0}{dr} + g_s \frac{d\phi_0}{dr}\right)\mathbf{l}\cdot\mathbf{S} \equiv -\alpha_{so}(r)\mathbf{l}\cdot\mathbf{S} \quad . \tag{2.28}$$

The maximum value α_{max} is compared in table 3 with $<\alpha_{so}(r)>$ obtained from the experimental spin-orbit splitting of the $f_{7/2} - f_{5/2}$ levels in Sc^{41}. Note that whereas the effects of V_0 and ϕ_0 cancel in the binding energy, they add in the spin-orbit interaction which, of course, forms the basis of the nuclear shell model.

The extension for nuclei with $N \neq Z$ is discussed by Walecka.[14] The calculation of the equation of state of nuclear matter at all temperatures as well as densities is presented in references 7 and 11.

Corrections to this mean-field theory coming from the effects of the mean meson fields on virtual baryon-antibaryon pairs can be taken into account through the addition of a "vacuum fluctuation" term, which provides a proper treatment of the vacuum energy remaining after extracting the normal-ordered part of the Hamiltonian in Eq. (2.14). This is carried out in references 8, 9, 11, 13, and 18. If the coupling constants are again required to fit the saturation properties of nuclear matter, the effect on the charge density for ${}_{20}Ca^{40}$, although essential to the mathematical consistency of the theory, is not large and is illustrated in figure 3. The inclusion of exchange corrections to the MFT is investigated in references 8, 9, 10, 11, and 13, and the effects of two-body correlations have been examined by Brittan.[10] Collective modes of the system in this model have been investigated by Chin,[8,9] and the thermal conductivity has been calculated recently by Freedman.[12]

III. NUCLEAR HYDRODYNAMICS

With this introduction, we proceed to the main purpose of the present paper, which is to calculate the flow properties of cold nuclear matter at various densities. The canonical stress tensor for the field theory in section II can be written as[5]

$$T_{\mu\nu} = \left\{ -\frac{1}{2}c^2\left[\left(\frac{\partial\phi}{\partial x_\lambda}\right)^2 + m_s^2\phi^2\right] - \frac{1}{4} F_{\lambda\varrho}F_{\lambda\varrho} - \frac{1}{2}m_v^2 V_\lambda V_\lambda\right\} \delta_{\mu\nu}$$

$$+ \hbar c\, \bar{\psi}\gamma_\mu \frac{\partial\psi}{\partial x_\nu} + c^2\left(\frac{\partial\phi}{\partial x_\mu}\right)\left(\frac{\partial\phi}{\partial x_\nu}\right) - \frac{\partial V_\lambda}{\partial x_\nu} F_{\lambda\mu} \quad . \tag{3.1}$$

This form is unsymmetrized, but that makes no difference for the present discussion of uniform systems. The stress tensor is conserved by construction:

$$\frac{\partial}{\partial x_\mu} T_{\mu\nu} = 0 \quad . \tag{3.2}$$

The four-momentum operator for the system is defined by

$$P_\mu = (\mathbf{P}, \frac{i}{c}\mathrm{H}) \tag{3.3a}$$

$$\equiv \frac{1}{ic}\int d\mathbf{x}\, T_{4\mu} \quad . \tag{3.3b}$$

Differentiation with respect to time and the use of Eq. (3.2) shows that P_μ is a constant of the motion. Furthermore, the integral in Eq. (3.3) is independent of the particular space-like hypersurface because of Eq. (3.2) and Gauss's theorem in four dimensions. Hence P_μ transforms like a Lorentz four-vector.

Consider now *uniform nuclear matter in a state of uniform flow*. We again make the mean-field approximation discussed in section II. The new features of the problem are a finite baryon current

$$\mathbf{B} = \text{const.} \tag{3.4}$$

and since there is now a direction in the problem, a finite spatial part to the vector field [cf. Eq. (2.8)]

$$<\mathbf{V}> \rightarrow \mathbf{V} = \text{const.} \tag{3.5}$$

These two quantities can be related through the use of the field equation (2.3), which implies

$$\mathbf{V} = \frac{g_v}{m_v^2}\mathbf{B} \quad . \tag{3.6}$$

This relation now augments Eqs. (2.9, 2.10). The required modifications of Eqs. (2.12) and (2.13) are

$$\epsilon_k^{(\pm)} = (g_v/\hbar c)\, V_0 \pm [(\mathbf{k} - (g_v/\hbar c)\mathbf{V})^2 + M^{*2}]^{1/2} \tag{3.7}$$

$$\bar{U}U = \frac{M^*}{[M^{*2} + (\mathbf{k} - (g_v/\hbar c)\mathbf{V})^2]^{1/2}}\, U^\dagger U \quad . \tag{3.8}$$

The Hamiltonian density and momentum density in the MFT follow from Eqs. (3.1, 3.3) as in Eq. (2.14).

$$\hat{\mathcal{H}}_{\text{MFT}} = g_v V_0 \varrho_B +$$

$$(\hbar c/\Omega)\, \Sigma_{\mathbf{k}\lambda}\, [(\mathbf{k} - g_v/\hbar c)\mathbf{V})^2 + M^{*2}]^{1/2}\; (A^\dagger_{\mathbf{k}\lambda}A_{\mathbf{k}\lambda} + B^\dagger_{\mathbf{k}\lambda}B_{\mathbf{k}\lambda})$$

$$+\frac{1}{2}m_s^2c^2\phi_0^2 - \frac{1}{2}m_v^2V_0^2 + \frac{1}{2}m_v^2\mathbf{V}^2 \tag{3.9}$$

$$\hat{\mathcal{P}}_{\text{MFT}} = (1/\Omega)\Sigma_{\mathbf{k}\lambda}\hbar\mathbf{k}(A^\dagger_{\mathbf{k}\lambda}A_{\mathbf{k}\lambda} + B^\dagger_{\mathbf{k}\lambda}B_{\mathbf{k}\lambda}) \quad . \tag{3.10}$$

Note the modification of the baryon energy spectrum in Eq. (3.9) and the presence of the final mass term for the field **V**.

Let us assume that the ground state of this uniform moving medium is obtained by filling the energy levels up to a Fermi surface $k_F = k_F(\theta_k,\phi_k)$ whose shape remains to be determined. In this case the energy, momentum, baryon, and scalar densities, and baryon current (calculated from Eq. (2.5) and the Dirac equation) become

$$\mathcal{E} = (\hbar c/\Omega)\Sigma_{\mathbf{k}_\lambda}^{k_F}((g_v/\hbar c)V_0 + [(\mathbf{k} - (g_v/\hbar c)\mathbf{V})^2 + M^{*2}]^{1/2})$$

$$+ \frac{1}{2}m_s^2c^2\phi_0^2 - \frac{1}{2}m_v^2V_0^2 + \frac{1}{2}m_v^2\mathbf{V}^2 \qquad (3.11)$$

$$\mathcal{P} = (1/\Omega)\,\Sigma_{\mathbf{k}_\lambda}^{k_F}\hbar\mathbf{k} \qquad (3.12)$$

$$\varrho_B = (1/\Omega)\,\Sigma_{\mathbf{k}_\lambda}^{k_F}1 \qquad (3.13)$$

$$\varrho_s = (1/\Omega)\,\Sigma_{\mathbf{k}_\lambda}^{k_F}\frac{M^*}{[M^{*2} + (\mathbf{k} - (g_v/\hbar c)\mathbf{V})^2]^{1/2}} \qquad (3.14)$$

$$\mathbf{B} = (1/\Omega)\,\Sigma_{\mathbf{k}_\lambda}^{k_F}\frac{(\mathbf{k} - (g_v/\hbar c)\mathbf{V})}{[M^{*2} + (\mathbf{k} - (g_v/\hbar c)\mathbf{V})^2]^{1/2}} \quad . \qquad (3.15)$$

The Fermi surface enters as an unknown function in these equations

$$\Sigma_{\mathbf{k}_\lambda}^{k_F} \xrightarrow[\Omega\to\infty]{} \frac{\Omega}{(2\pi)^3}\Sigma\lambda\int\int \sin\theta_k d\theta_k d\phi_k \int_0^{k_F(\theta_k,\phi_k)} k^2dk \quad . \qquad (3.16)$$

The shape of this Fermi surface may be obtained by using the thermodynamic argument that at a *fixed* baryon density ϱ_B and momentum density $\mathcal{P}$ (recall that both these quantities are constants of the motion), the system in equilibrium will minimize the energy density. We may speak of densities, since for a uniform system the integral quantities involve only multiplication by the fixed volume Ω. This is now a straightforward variational problem. We minimize $\mathcal{E}$ and incorporate the constraints through Lagrange multipliers μ and $\mathbf{v}$ [cf. Eq. (2.22)]

$$\delta\mathcal{E}(k_F; \phi_0, V_0, \mathbf{V}) - \mu\delta\varrho_B(k_F) - \mathbf{v}\cdot\delta\mathcal{P}(k_F) = 0 \quad . \qquad (3.17)$$

Here the additional explicit dependence of $\mathcal{E}$ on the variables $\{\phi_0, V_0, \mathbf{V}\}$ has been noted; $\delta\varrho_B$ and $\mathcal{P}$ contain no explicit dependence on these quantities. The calculation is greatly simplified when it is realized that the mean field equations (2.9, 2.10, and 3.6) ensure that the variations of the energy density with respect to these quantities vanish.

$$\left(\frac{\delta \mathcal{E}}{\delta \phi_0}\right)_{k_F; V_0, \mathbf{V}} = \left(\frac{\delta \mathcal{E}}{\delta V_0}\right)_{k_F; \phi_0, \mathbf{V}}$$

$$= \left(\frac{\delta \mathcal{E}}{\delta \mathbf{V}}\right)_{k_F; \phi_0, V_0} = 0 \quad . \tag{3.18}$$

Thus they may be held constant during the variation of the surface $\delta k_F(\theta_k, \phi_k)$. The remaining calculation using Eq. (3.16) is elementary and yields for the equation of the Fermi surface

$$g_v V_0 + \hbar c[M^{*2} + (\mathbf{k}_F - (g_v/\hbar c)\mathbf{V})^2]^{1/2} - \hbar \mathbf{v} \cdot \mathbf{k}_F = \mu = \text{const.} \tag{3.19}$$

It is convenient now to introduce a new variable

$$\mathbf{t} \equiv \mathbf{k} - (g_v/\hbar c)\mathbf{V} \tag{3.20}$$

and again introduce dimensionless variables as in section II. Equations (3.11, 3.12, 3.13, and 3.15) can then be put in the form

$$e = \frac{1}{2} C_v^2 n_B^2 + \frac{1}{2C_s^2}(1 - \chi)^2 + \frac{1}{2} C_v^2 \mathbf{b}^2 +$$

$$\frac{\gamma}{(2\pi)^3} \int_0^{t_F} d\mathbf{t}(t^2 + \chi^2)^{1/2} \tag{3.21}$$

$$\mathbf{p} = C_v^2 n_B \mathbf{b} + \frac{\gamma}{(2\pi)^3} \int_0^{t_F} \mathbf{t} \, d\mathbf{t} \tag{3.22}$$

$$n_B = \frac{\gamma}{(2\pi)^3} \int_0^{t_F} d\mathbf{t} \tag{3.23}$$

$$\mathbf{b} = \frac{\gamma}{(2\pi)^3} \int_0^{t_F} \frac{\mathbf{t}}{(t^2 + \chi^2)^{1/2}} d\mathbf{t} \tag{3.24}$$

where $\mathbf{b}$ is now the dimensionless baryon current. The equation for the Fermi surface (3.19) becomes

$$(\chi^2 + \mathbf{t}_F^2)^{1/2} - \mathbf{v} \cdot \mathbf{t}_F = \mu_{\text{eff}} \tag{3.25}$$

with

$$\mu_{\text{eff}} \equiv (\mu/m_b c^2) - C_v^2 n_B(1 - \mathbf{v} \cdot \mathbf{b}/n_B) \tag{3.26}$$

and $\mathbf{v} \equiv \mathbf{v}/c$ in dimensionless units. The self-consistency equation for the effective mass χ can be obtained from the minimization of the energy density with respect to that parameter.

$$\chi = [1 + C_s^2 \frac{\gamma}{(2\pi)^3} \int_0^{t_F} d\mathbf{t} \frac{1}{(t^2 + \chi^2)^{1/2}}]^{-1} \quad . \tag{3.27}$$

This is identical to the scalar meson field equation (2.9) where the right-hand side has been calculated with Eq. (3.14). The previous results for uniform nuclear matter at rest in section II can be recovered simply by setting $\mathbf{v} = 0$ in the expressions above, in which case the Fermi surface defined in Eq. (3.25) becomes spherically symmetric, implying $\mathbf{b} = 0$ and then $\mathcal{P} = 0$ from Eqs. (3.24, 3.22).

To examine the structure of these equations consider first the *non-relativistic limit* (NRL) where

$$(\chi^2 + t^2)^{1/2} \cong \chi + t^2/2\chi \quad . \tag{3.28}$$

A simple change of variables

$$\mathbf{t} \equiv \mathbf{s} + \chi\mathbf{v} \tag{3.29}$$

with $d\mathbf{t} = d\mathbf{s}$ then reduces Eq. (3.25) to

$$s_F^2/2\chi = \mu_{\text{eff}} - \chi + \frac{1}{2}\chi v^2 \tag{3.30}$$

and the Fermi surface is spherically symmetric in the variable s. Equation (3.23) then yields

$$n_B = \frac{\gamma}{6\pi^2} s_F^3 \tag{3.31}$$

which expresses s_F in terms of the baryon density. The baryon current is evaluated from Eq. (3.24) in this limit to give

$$\mathbf{b} \cong n_B \mathbf{v} (1 - \frac{1}{2\chi^2} \left(\frac{6\pi^2 n_B}{\gamma} \right)^{2/3}) \tag{3.32}$$

where we here work to lowest order in v assuming $v^2 << 1$. We next define the *hydrodynamic flow velocity* through the relation

$$\mathbf{b} \equiv n_B \mathbf{v}_{\rm hyd} \quad . \tag{3.33}$$

Equation (3.32) therefore relates the Lagrange multiplier $\mathbf{v}$ to the flow velocity $\mathbf{v}_{\rm hyd}$ according to

$$\mathbf{v}_{\rm hyd} \cong \mathbf{v}\left(1 - \frac{1}{2\chi^2}\left(\frac{6\pi^2 n_B}{\gamma}\right)^{2/3}\right) \quad . \tag{3.34}$$

The solution to the self-consistency Eq. (3.27) in this non-relativistic limit is

$$\chi \cong 1 - C_s^2 n_B \quad . \tag{3.35}$$

The momentum density is obtained from Eq. (3.22) as

$$\mathscr{P} = n_B C_v^2 \mathbf{b} + \frac{\gamma}{(2\pi)^3} \int_0^{t_F} \mathbf{t}\, dt \quad . \tag{3.36}$$

We may use this relation to define the *mass of the hydrodynamic flow* according to

$$\mathscr{P} = \chi_{\rm hyd}\, \mathbf{b} \tag{3.37a}$$

$$\mathscr{P} = \chi_{\rm hyd}\, n_B\, \mathbf{v}_{\rm hyd} \tag{3.37b}$$

and a combination of Eqs. (3.34, 3.35, 3.36) then yields

$$\chi_{\rm hyd} \cong 1 + \frac{1}{2}\left(\frac{6\pi^2 n_B}{\gamma}\right)^{2/3} + n_B(C_v^2 - C_s^2) \qquad ;\text{NRL} \tag{3.38}$$

as the non-relativistic limit of the hydrodynamic mass. Note that the formal expansion parameter is the baryon density [Eq. (3.31)].

The energy density of the system given by Eq. (3.21) differs only by terms of order v^2 from that of nuclear matter at rest at the same baryon density, which can be written in this limit as

$$e \cong n_B + \frac{3}{10}\left(\frac{6\pi^2 n_B}{\gamma}\right)^{2/3} n_B + \frac{1}{2}\, n_B^2(C_v^2 - C_s^2) \qquad ;\text{NRL.} \tag{3.39}$$

We observe from Eqs. (3.38) and (3.39) that the hydrodynamic mass in this non-relativistic limit can be written as

$$\chi_{\text{hyd}} = \left(\frac{\partial e}{\partial n_B} \right)_0 \qquad ;\text{NRL} \qquad (3.40)$$

where the subscript zero denotes the rest frame of the fluid. The pressure of nuclear matter at rest at the baryon density n_B is obtained from Eq. (2.16) as

$$p \cong \frac{1}{5}\left(\frac{6\pi^2 n_B}{\gamma} \right)^{2/3} n_B + \frac{1}{2} n_B^2 (C_v^2 - C_s^2) \qquad ;\text{NRL}. \qquad (3.41)$$

Thus an alternative expression for the hydrodynamic mass is

$$\chi_{\text{hyd}} = \left(\frac{e + p}{n_B} \right)_0 \qquad ;\text{NRL}. \qquad (3.42)$$

Consider next the *extreme relativistic limit* (ERL) of this theory where $\chi \to 0$ and v is arbitrary (with $v < 1$). In this case the Fermi surface is given by Eq. (3.25) as

$$t_F - \mathbf{v} \cdot \mathbf{t}_F = \mu_{\text{eff}} \quad . \qquad (3.43)$$

If we choose $\mathbf{v}$ as the z-axis in momentum space, then the Fermi surface becomes

$$t_F = \mu_{\text{eff}}/(1 - v \cos \theta_t) \quad . \qquad (3.44)$$

This is the equation of an ellipse. In the present dimensionless units Eq. (3.20) implies $\mathbf{k} = \mathbf{t} + n_B C_v^2 \mathbf{v}$ and hence the focus of the ellipse is at the point $n_B C_v^2 \mathbf{v}$ in momentum space. The semi-minor axis is given by $\mu_{\text{eff}}/(1 - v^2)^{1/2}$ and semi-major axis $\mu_{\text{eff}}/(1 - v^2)$. The baryon density is obtained from Eq. (3.23) as [cf. Eq. (3.16)]

$$n_B = \frac{\gamma}{6\pi^2} \frac{\mu_{\text{eff}}^3}{(1 - v^2)^2} \qquad (3.45)$$

which expresses μ_{eff} in terms of n_B at any v. The baryon current in Eq. (3.24) is

$$\mathbf{b} \cong n_B \mathbf{v} \qquad (3.46)$$

which identifies the Lagrange multiplier $\mathbf{v}$ as the hydrodynamic flow velocity defined through Eq. (3.33). The momentum density follows from Eq. (3.22):

$$\mathbf{p} = n_B C_v^2 \mathbf{b} + n_B \mathbf{v} \frac{\mu_{\text{eff}}}{(1 - v^2)} \quad . \tag{3.47}$$

We again define the mass for hydrodynamic flow according to Eqs. (3.37) and hence can identify in this limit

$$\chi_{\text{hyd}} \cong n_B C_v^2 + \left(\frac{6\pi^2 n_B}{\gamma} \frac{1}{1 - v^2} \right)^{1/3} \quad ; \text{ERL.} \tag{3.48}$$

Note in this case we are discussing the high-density expansion of the hydrodynamic mass. It is just in this high-density regime that the present mean field theory is expected to become most valid. The solution to the self-consistency equation (3.27) in this limit is

$$\chi \cong [1 + C_s^2 \left(\frac{9\gamma}{16\pi^2} n_B^2 (1 - v^2) \right)^{1/3}]^{-1} \tag{3.49}$$

and indeed $\chi \rightarrow 0$ as $n_B \rightarrow \infty$ for any v, verifying our initial ERL assumption.

Let us now try to relate the quantities above to corresponding quantities in the rest frame of the fluid (denoted with a superscript zero). First, since the number of baryons dN_{baryons} in a given element $d\mathbf{x}^0$ of fluid is unchanged when the fluid is set into motion, we have (recall all lengths are in units of $1/M$)

$$dN_{\text{baryons}} = n_B^0 d\mathbf{x}^0 = n_B d\mathbf{x} \quad . \tag{3.50}$$

The Lorentz contraction of the longitudinal dimension of the volume tells us that $d\mathbf{x} = d\mathbf{x}^0(1 - v^2)^{1/2}$ and hence the baryon density in the rest frame is related to the baryon density of the moving fluid by

$$n_B = n_B^0/(1 - v^2)^{1/2} \quad . \tag{3.51}$$

The hydrodynamic mass in Eq. (3.48) can thus be rewritten as

$$\chi_{\text{hyd}} = \chi_{\text{hyd}}^0/(1 - v^2)^{1/2} \tag{3.52a}$$

where the hydrodynamic rest mass is defined by

$$\chi^0_{\text{hyd}} \equiv n_B^0 C_v^2 + \left(\frac{6\pi^2 n_B^0}{\gamma} \right)^{1/3} \qquad ;\text{ERL.} \qquad (3.52\text{b})$$

It is evident from Eq. (3.52a) that this mass shows the correct relativistic increase with velocity. The energy and pressure of nuclear matter at rest with density n_B^0 are given by Eqs. (2.15, 2.16)

$$e^0 \cong \frac{1}{2} C_v^2 (n_B^0)^2 + \frac{3}{4} \left(\frac{6\pi^2 n_B^0}{\gamma} \right)^{1/3} n_B^0 \qquad (3.53\text{a})$$

$$p^0 \cong \frac{1}{2} C_v^2 (n_B^0)^2 + \frac{1}{4} \left(\frac{6\pi^2 n_B^0}{\gamma} \right)^{1/3} n_B^0 \qquad ;\text{ERL.} \qquad (3.53\text{b})$$

Thus the hydrodynamic mass in Eq. (3.52b) can be expressed either as [cf. Eq. (3.40)]

$$\chi^0_{\text{hyd}} = \left(\frac{\partial e}{\partial n_B} \right)_0 \qquad ;\text{ERL} \qquad (3.54)$$

or as [cf. Eq. (3.42)]

$$\chi^0_{\text{hyd}} = \left(\frac{e+p}{n_B} \right)_0 \qquad ;\text{ERL.} \qquad (3.55)$$

Equations (3.48) and (3.52) are the main results of the present work. The hydrodynamic mass gets very large, and hence the medium gets very stiff, either for high density ($n_B \to \infty$, any v) or high velocity (fixed large n_B^0, $v \to 1$). This is the mass that enters into Newton's second law for the rate of change of momentum and the basic equations of hydrodynamics for the system.

We may understand Eq. (3.54) in the following manner: Suppose a few baryons are added to a volume $d\mathbf{x}^0$ of nuclear matter at rest so that the density in this region is changed by δn_B^0 and the energy density by

$$\delta e^0 = \left(\frac{\partial e}{\partial n_B} \right)_0 \delta n_B^0 \quad . \qquad (3.56)$$

The change in energy of the system $\delta\epsilon^0$ is given as

$$\delta\epsilon^0 = \delta e^0 d\mathbf{x}^0 \quad . \tag{3.57}$$

Now the energy and momentum of the system defined by Eqs. (3.3) form a four-vector

$$\delta p_\mu^0 = (\mathbf{0}, i\delta\epsilon^0) \tag{3.58}$$

where we henceforth denote the dimensionless stress tensor and four-momentum by $t_{\mu\nu}$ and $p_\mu \equiv (\mathbf{p}, i\epsilon)$ respectively. Thus for a Lorentz transformation in the z-direction (along $\mathbf{v}$)

$$a_{\mu\nu}(v) = \begin{bmatrix} 1 & 0 & 0 & 0 \\ 0 & 1 & 0 & 0 \\ 0 & 0 & \dfrac{1}{(1-v^2)^{1/2}} & \dfrac{-iv}{(1-v^2)^{1/2}} \\ 0 & 0 & \dfrac{iv}{(1-v^2)^{1/2}} & \dfrac{1}{(1-v^2)^{1/2}} \end{bmatrix} \tag{3.59}$$

after which the fluid appears to be moving with velocity v in the z-direction, the Lorentz transformed four-momentum

$$\delta p_\mu = a_{\mu\nu}\delta p_\nu^0 \tag{3.60}$$

is given with the aid of Eq. (3.50) by

$$\begin{aligned} \delta\mathbf{p} &= \frac{\mathbf{v}}{(1-v^2)^{1/2}}[\delta e^0 d\mathbf{x}^0] \\ &= \frac{\mathbf{v}}{(1-v^2)^{1/2}}\left(\frac{\partial e}{\partial n_B}\right)_0 \delta n_B d\mathbf{x} \quad . \end{aligned} \tag{3.61}$$

Therefore the additional momentum density in the moving fluid is obtained as

$$\delta \not{p} = \frac{\mathbf{v}}{(1-v^2)^{1/2}} \left(\frac{\partial e}{\partial n_B} \right)_0 \delta n_B \tag{3.62}$$

and we recover our previous result for χ_{hyd} in Eq. (3.54).

Equation (3.55) may be understood as follows. The stress tensor in Eq. (3.1) forms a second-rank tensor. For a uniform fluid at rest, the stress tensor in Eq. (1.1) takes the form

$$t^0_{\mu\nu} = p\delta_{\mu\nu} + (e+p)\, u_\mu u_\nu \tag{3.63}$$

with $u_\mu = (\mathbf{0}, ic)$. After the Lorentz transformation (3.59), the stress tensor for the moving fluid is

$$t_{\mu\nu} = a_{\mu\mu'}\, a_{\nu\nu'}\, t^0_{\mu'\nu'} \quad . \tag{3.64}$$

Since the medium is uniform before and after the Lorentz transformation, we need not worry about the point at which the tensor is evaluated. Explicit evaluation of the sum in Eq. (3.64) using Eqs. (3.59) and (3.63) yields

$$t_{4j} = \frac{iv_j}{(1-v^2)} (p+e)_0 \quad . \tag{3.65}$$

The momentum is given by Eq. (3.3) as

$$p_j = \frac{1}{i} \int d\mathbf{x} t_{4j} \tag{3.66}$$

or equivalently, with the aid of Eq. (3.51)

$$\mathbf{p} = \int \left[\frac{\mathbf{v}}{(1-v^2)^{1/2}} \left(\frac{p+e}{n_B} \right)_0 \right] n_B d\mathbf{x} \quad . \tag{3.67}$$

Thus the momentum density is

$$\not{p} = \frac{\mathbf{v}}{(1-v^2)^{1/2}} \left(\frac{p+e}{n_B} \right)_0 n_B \tag{3.68}$$

and we recover Eq. (3.55).

The equivalence of Eqs. (3.54) and (3.55) is just thermodynamics, for if the number of baryons B is kept constant and the volume Ω is varied, we have

$$\frac{\partial e}{\partial n_B} = \frac{\partial(\epsilon/\Omega)}{\partial(B/\Omega)} = -\frac{\Omega^2}{B}\frac{\partial}{\partial\Omega}\left(\frac{\epsilon}{\Omega}\right)$$

$$= -\frac{1}{n_B}\left(\frac{\partial\epsilon}{\partial\Omega}\right)_B + \frac{e}{n_B} = \frac{p+e}{n_B} \tag{3.69}$$

by the definition of the pressure as $p = -(\partial\epsilon/\partial\Omega)_B$.

In summary, we have developed a general analysis for the uniform flow of nuclear matter in this mean field theory, and we have shown that the mass for hydrodynamic flow is given at high densities by Eqs. (3.48) and (3.52). One can apply the present results to non-uniform flow with the same approach as used in the discussion of finite nuclei in section II. Although the present field theory severely oversimplifies the actual meson degrees of freedom in the nucleus, it is interesting to have a model that (1) describes observed nuclear matter, (2) never introduces the concept of a static two-nucleon potential, (3) explicitly takes into account the meson degrees of freedom, and (4) exhibits Lorentz covariance at every stage. The model thus, at least in principle, allows an extrapolation away from ordinary nuclear matter to regimes of very high density and very large flow velocities.

ACKNOWLEDGMENT

This work was supported in part by NSF Grant No. NSF PHY 77-16188.

REFERENCES

1. F. Bloch, Z. Phys. **81**, 363 (1933).
2. J. D. Walecka, Phys. Lett. **58A**, 83 (1976).
3. J. D. Walecka, Phys. Lett. **61A**, 230 (1977).
4. F. E. Serr, Phys. Lett. **62A**, 325 (1977).
5. J. D. Walecka, Ann. Phys. (NY) **83**, 491 (1974).
6. S. A. Chin and J. D. Walecka, Phys. Lett. **52B**, 24 (1974).
7. J. D. Walecka, Phys. Lett. **59B**, 109 (1975).
8. S. A. Chin, Phys. Lett. **62B**, 263 (1976).
9. S. A. Chin, Ann. Phys. (NY) **108**, 301 (1977).
10. M. Brittan, Phys. Lett. **79B**, 27 (1978).
11. R. A. Freedman, Phys. Lett. **71B**, 369 (1977).

12. R. A. Freedman, Lett. Nuovo Cimento **25**, 399 (1979).
13. F. E. Serr and J. D. Walecka, Phys. Lett. **79B**, 10 (1978).
14. J. D. Walecka, Lecture Notes in Physics **92**, Nuclear Interactions (Berlin, Springer-Verlag, 1978), p. 294.
15. M. H. Johnson and E. Teller, Phys. Rev. **98**, 783 (1955).
16. H. P. Duerr, Phys. Rev. **103**, 469 (1956).
17. T. D. Lee and G. C. Wick, Phys. Rev. **D9**, 2291 (1974).
18. T. D. Lee and M. Margulis, Phys. Rev. **D11**, 1591 (1975).
19. A. K. Kerman and L. D. Miller, Center for Theoretical Physics Pub. #449, M.I.T. (1974).
20. J. Boguta and A. R. Bodmer, Nucl. Phys. **A292**, 413 (1977).
21. J. Boguta and J. Rafelski, Phys. Lett. **71B**, 22 (1977).
22. A. L. Fetter and J. D. Walecka, Quantum Theory of Many-Particle Systems (New York, McGraw-Hill Book Co., 1971).
23. Ya. B. Zel'dovich, Sov. Phys.—JETP **14**, 1143 (1962).

POINTWISE SO_4 SYMMETRY OF THE BPST PSEUDOPARTICLE SOLUTION

by Chen Ning Yang

ABSTRACT

The BPST pseudoparticle solution is shown to be everywhere pointwise SO_4 symmetrical. It is further shown that on flat Euclidean space R_4, the only SU_2 gauge field that is everywhere SO_4 symmetrical is a BPST pseudoparticle solution.

I. INTRODUCTION

In a recent generalization[1] of the Dirac monopole to SU_2 gauge fields, it was found that the concept of "pointwise SO_4 symmetry" is useful. The meaning of this concept[2] can be explained in the following way:

Consider a four-dimensional manifold with a Riemannian geometry having + + + + signature. Let P be a point on the manifold. Consider an SU_2 gauge field with field strengths $(f^i_{\mu\nu})_P$ at the point P. Does $(f^i_{\mu\nu})_P$ serve to differentiate between the various directions from P? If it does not, we say the field has pointwise SO_4 symmetry at P. To be more precise, choose coordinates so that the metric at P is $g_{\mu\nu}=\delta_{\mu\nu}$. If any SO_4 rotation for the indices μ,ν in $(f^i_{\mu\nu})_P$ can be *compensated* for by a gauge transformation on the index i, the field has pointwise SO_4 symmetry at P.

In this paper I show that the pseudoparticle solution[3] of Belavin, Polyakov, Schwartz, and Tynpkin, to be called the BPST solution, is everywhere pointwise SO_4 symmetrical. I then show that the only gauge field (sourceless or not) on R_4, (i.e., on flat + + + + space), that is everywhere pointwise SO_4 symmetrical is the BPST solution.

Chen Ning Yang is Director of the Institute of Theoretical Physics and Professor of Physics at the State University of New York, Stony Brook.

II. POINTWISE SO_4 SYMMETRY OF BPST SOLUTION

It was shown in the appendix of reference 1 that the following statements are identical

(a) $f^i_{\mu\nu}$ is pointwise SO_4 symmetrical at P, (also called orthogonal or regular at P),
(b) $f^i_{\mu\nu} f^{j\lambda\nu} = a^2 \delta^{ij} \delta^\lambda_\mu + a\epsilon^{ijk} f^{k\cdot\lambda}_\mu$ at P,
(c) $f^i_{\mu\nu} f^{j\lambda\nu} + f^j_{\mu\nu} f^{i\lambda\nu} = 2a^2 \delta^{ij} \delta^\lambda_\mu$ at P,

where a is a scalar function on the manifold. It is further easy to show from lemmas 1α, 1β and 4 of reference 1 that these statements are also identical to

(d) $f^i_{\mu\nu}$ is self dual or self antidual at P, and in a coordinate system for which $g_{\mu\nu} = \delta_{\mu\nu}$ at P,

$$\tilde{\mathcal{E}}\,\mathcal{E} = \tilde{\mathcal{H}}\,\mathcal{H} = a^2 \quad .$$

Theorem 1. The BPST solution is everywhere pointwise SO_4 symmetrical.

Proof: By a straightforward evaluation of the field strengths $\mathcal{E}$ and $\mathcal{H}$ for the BPST solution, we easily verify property (d) above. Hence the theorem is proved.

The square of the field strength, a^2, is easily computed to be

$$a^2 = \frac{16K^2}{[x^2 + K]^4} \quad (K > 0).$$

where $x^2 = x_1^2 + x_2^2 + x_3^2 + x_4^2$ and x_μ are the Cartesian coordinates. This function peaks at $x = 0$. Around a point P where $x \neq 0$, a SO_4 rotation of the whole field changes the magnitude of the field strength squared at most points. So the field is not SO_4 symmetrical at P. But it is pointwise SO_4 symmetrical at P in that if one considers only the value of $f^i_{\mu\nu}$ at P, then the rotation of the field strengths is equivalent to a gauge transformation of the original field strengths. Thus the value of $f^i_{\mu\nu}$ at P does not serve to choose an SO_4 frame around P.

Theorem 2. The BPST solution is the only SU_2 gauge field that is everywhere pointwise SO_4 symmetrical on R_4, the flat 4-dimensional Euclidean space.

Proof: (1) According to Appendix A of reference 1, a field that is pointwise SO_4 symmetrical can be gauge transformed to the standard form, with only one parameter, a, its amplitude. For a field that is everywhere pointwise SO_4 symmetrical, the field strengths in the proper gauge are of the standard form (IA10) or (IA11) everywhere. The amplitude a is a function of the x's. We shall write these equations in the following form

$$f^i_{\mu\nu} = a\eta^i_{\mu\nu} \quad , \tag{1}$$

or

$$f^i_{\mu\nu} = a\bar{\eta}^i_{\mu\nu} \quad , \tag{2}$$

where η and $\bar{\eta}$ are the symbols introduced by 't Hooft[4]:

$$\eta^i_{\mu\nu} = \epsilon_{i\mu\nu4} + \delta_{\mu i}\delta_{\nu 4} - \delta_{\nu i}\delta_{\mu 4} \quad , \tag{3}$$

$$\bar{\eta}^i_{\mu\nu} = \epsilon_{i\mu\nu4} - \delta_{\mu i}\delta_{\nu 4} + \delta_{\nu i}\delta_{\mu 4} \quad . \tag{4}$$

For the self-dual case (1), the Bianchi identity becomes[5]

$$\eta^{\xi\mu\nu\lambda}(\eta^i_{\mu\nu}a_{,\lambda} - aC^i_{jk}\eta^j_{\mu\nu}b^k_\lambda) = 0 \quad . \tag{5}$$

Since $\eta^i_{\mu\nu}$ is self dual, this becomes

$$\eta^{i\xi\lambda}a_{,\lambda} - aC^i_{jk}\,\eta^{j\xi\lambda}b^k_\lambda = 0 \quad . \tag{6}$$

(2) Now define a matrix M and columns Δ and b by

$$<i\xi|M|k\lambda> = C^i_{jk}\,\eta^{j\xi\lambda} \quad , \tag{7}$$

$$<i\xi|\Delta> = \eta^{i\xi\lambda}a_{,\lambda} \quad , \tag{8}$$

$$<k\lambda|b> = b^k_\lambda \quad . \tag{9}$$

Eq. (6) becomes

$$\Delta - aMb = 0 \quad . \tag{10}$$

Using[4,1]

$$\eta^i_{\alpha\lambda}\eta^j_{\beta\lambda} = \epsilon^{ijk}\,\eta_{\alpha\beta} + \delta^{ij}\delta_{\alpha\beta} \tag{11}$$

we can prove

$$(M + 1)M = 2 \quad . \tag{12}$$

Thus

$$b = a^{-1}(M + 1)\Delta/2 \quad .$$

Or

$$b^i_\xi = -a_{,\mu}\eta\xi_\mu(2a)^{-1} \quad . \tag{13}$$

(3) Substituting this into the equation for $f^i_{\mu\nu}$ in terms of b^i_ξ and its derivatives, we obtain as necessary and sufficient conditions for (1):

$$A_{,\mu\nu} = \delta_{\mu\nu}B + A_{,\mu}A_{,\nu} \quad , \tag{14}$$

$$2a = -A_{,\mu\mu} - (A_{,\mu})^2 \quad , \tag{15}$$

where

$$A = \frac{1}{2}\,ln|a| \quad , \tag{16}$$

and B is a scalar function of the coordinates. Putting

$$G = \exp(-A) \quad , \tag{17}$$

(14) becomes

$$G_{,ij} = 0,(i \neq j) \quad ,$$

and

$$G_{,11} = G_{,22} = G_{,33} = G_{,44} \quad .$$

These equations can be integrated, giving

$$G = \alpha(x - c)^2 + \beta \tag{18}$$

where c_μ is a point in R_4, and α and β are numbers. Substitution into (10), (16), and (17) gives

$$a = \frac{\pm 1}{(\alpha(x - c)^2 + \beta)^2} \quad .$$

To avoid singularities in $f^i_{\mu\nu}$, a must remain finite. Thus $\alpha\beta \nless 0$. Eq. (15) gives then $4\alpha\beta = 1$, and we obtain, with $K = \beta\alpha^{-1} > 0$,

$$a = \frac{4K}{[(x-c)^2 + K]^2} \quad , \tag{19}$$

and

$$b^i_\xi = -a_{,\mu}\, \eta^i_{\xi\mu}(2a)^{-1} \quad . \tag{20}$$

These two equations give exactly the BPST solution.[3] (20) is precisely in the form of the Corrigan-Fairlie-Wilczek-'t Hooft Ansatz.[6]

(4) The proof for the antiself dual case is entirely similar.

ACKNOWLEDGMENT

This work is supported in part by the National Science Foundation under Grant No. PHY-76-15328.

REFERENCES

1. Chen Ning Yang, J. Math. Phys. **19**, 320 (1978). Formulas in this paper will be referred to as (I2), (IA11), etc.
2. In reference 1, this concept was applied only to points P where the underlying geometry is SO_4 symmetrical at the point P. The concept is actually applicable to any point on any Riemannian geometry of signature + + + +.
3. Belavin, Polyakov, Schwartz, and Tynpkin, Phys. Lett. **59B**, 85 (1975).
4. G. 't Hooft, Phys. Rev. **D14**, 3432 (1976).
5. We use the notation that $\eta^{\xi\mu\nu\lambda} = \sqrt{g}\epsilon^{\xi\mu\nu\lambda}$ where $\epsilon = \pm 1$ is the antisymmetrical tensor. All notations follow that of reference 1.
6. E. Corrigan and D. B. Fairlie, Phys. Lett. **67B**, 69 (1977); F. Wilczek, in *Quark Confinement and Field Theory*, ed. by D. Stump and D. Weingarten (Wiley, New York, 1977); 't Hooft, unpublished.